HOMELAND SECURITY

HOMELAND SECURITY

What Is It and Where Are We Going?

Amos N. Guiora

CRC Press
Taylor & Francis Group
Boca Raton   London   New York

CRC Press is an imprint of the
Taylor & Francis Group, an **informa** business

CRC Press
Taylor & Francis Group
6000 Broken Sound Parkway NW, Suite 300
Boca Raton, FL 33487-2742

Printed in the United States of America on acid-free paper
10 9 8 7 6 5 4 3 2 1

International Standard Book Number: 978-1-4398-3818-1 (Hardback)

Library of Congress Cataloging-in-Publication Data

Guiora, Amos N., 1957-
    Homeland Security : what is it and where are we going? / Amos N. Guiora.
       p. cm.
    Includes bibliographical references and index.
    ISBN 978-1-4398-3818-1 (hardback)
    1. United States. Dept. of Homeland Security. 2. United States. Dept. of Homeland Security--Appropriations and expenditures. 3. Terrorism--United States--Prevention. 4. National security--United States. 5. Civil defense--United States. 6. Emergency management--United States. I. Title.

    HV6432.4.G85 2011
    363.340973--dc22   2011009632

Visit the Taylor & Francis Web site at
http://www.taylorandfrancis.com

and the CRC Press Web site at
http://www.crcpress.com

A person is, in many ways, judged by his friends. In that regard, I have been incredibly fortunate to have lifelong friends who I call the Brown Jug Gang—Lisa (Oneal) Conway, Rob Crim, Blake Roessler, John Lentz, and Gerry Korngold, who have truly been a critical part of my life journey, ranging from elementary school (Lisa) to high school (Rob) to college (Blake and John) to academia (Gerry). How lucky I am.

CONTENTS

PREFACE

Writing a book on homeland security is a daunting undertaking given the range, size, and complexity of the subject, particularly when the term itself is, largely, undefined. I come to the book with two distinct and significant homeland security "notches on my belt": I served for 5 years as the judge advocate for the Israel Defense Forces Home Front Command (the American version of the Department of Homeland Security) and was the legal advisor to a U.S. Congress task force (Committee on Homeland Security) mandated to develop U.S. homeland security strategy, which included testimony before the Congress. The upshot of the two professional experiences is significant familiarity with the issue coupled with deep skepticism regarding creation of an entity called homeland security.

Familiarity for obvious reasons; skepticism because I believe the issue is particularly unwieldy. Creating Home Front Command (HFC) and Department of Homeland Security (DHS) always struck me as reflecting political considerations seeking to mollify a concerned public in the aftermath of attacks. In the Israeli paradigm, HFC was established in the aftermath of the first Iraq war, when Israel was attacked by Iraqi SCUD missiles and Israelis spent significant time in bomb shelters and "protected" rooms in their homes; in the American paradigm, DHS was created in the aftermath of 9/11.

That lack of reflection and deep thinking was obvious to me when I undertook both positions and was reinforced to me on—literally—a daily basis. However, that lack of sophisticated thinking in no way minimizes the significance of both entities; after all, homeland security is of extraordinary importance to the civilian population. That is perhaps the irony in both paradigms: decision makers created—perhaps cynically—two enormous infrastructures to placate respective publics that if managed correctly with carefully prescribed "mission articulations" can actually benefit the public. However, as will be discussed throughout this book, the lack of careful and thoughtful definitions regarding *just what is* homeland security raises enormous stumbling blocks to ensuring effective homeland security.

The reality of developing—much less implementing—a homeland security infrastructure in the two countries was the direct result of political necessity. While necessity is the mother of all inventions, the lack of careful attention to the product's form is particularly problematic, both because of the enormous resources allocated to an undefined entity and because of the tragic reality surrounding its birth. Theoretically, homeland security makes sense; after all, government is obligated to protect those who live within its borders. While disagreement among reasonable people regarding the extent of that protection is legitimate, the basic duty is, largely, an agreed-upon consensus. That is not to gainsay the validity of the "how much/who pays/who gets" debate regarding government services; it is, however, to state the obvious regarding government duty with respect to safety and security.

Nevertheless, the creation of the two infrastructures was done in large part, devoid of sophisticated analysis regarding, among other issues:

- what needs homeland security is expected to address
- how to determine/define effectiveness
- how to determine resource allocation and prioritization
- what are the costs-benefits in creating homeland security infrastructures
- how to determine/resolve lines of authority with other competing bureaucracies
- how to define risks and threats enabling sophisticated risk and threat assessment

On the other hand, in allocating significant resources, both governments expected to develop a working infrastructure; in reality, both models were largely devoid of sufficient instruction and direction to develop strategy. While tactical decisions could be made, long-term issues reflective of more thoughtful decision making could not be addressed, as core questions were not resolved.

In undertaking this book I am acutely aware of the pitfalls and pratfalls inherent in discussing an issue as broad, complex, and largely undefined as homeland security. While, hopefully, my practical hands-on experience will be deemed helpful by the reader, I am under no illusion regarding the ability of one book to satisfactorily and honestly address enormous myriad issues. That said, it is my hope that those experiences combined with significant research, thinking, and writing on homeland security issues will assist the reader in developing a more informed opinion regarding homeland security.

The term *road map* is often used to describe undertakings of this nature; for a variety of reasons, I reject that term in this context. My efforts are intended to educate the reader regarding a variety of homeland security issues, based in part on concrete examples, some drawn on personal experiences, others not. In drawing on my personal experiences, the book includes examples from the United States and Israel alike; while this is not a comparative book, a broader perspective can be effective and productive.

To that end, the book seeks to shed light on specific issues; while I am critical of policies articulated, created, and implemented, this book is not intended to bash decisions or decision makers. It is, indeed, offered as a realistic assessment of homeland security—where it is and where we are going.

When Mark Listewnik first approached me proposing this project, I was initially skeptical both because of preexisting contractual obligations with respect to other book projects and because I felt others had—to varying degrees of success and interest—written on the subject. Mark's initial suggestion was that I focus on money laundering—an issue I had previously addressed; as important as that issue is (see Chapter 7), I did not feel that a book exclusively dedicated to one aspect of homeland security was something I wanted to undertake. However, as we continued our conversations over a period of time, I increasingly became convinced that my unusual perch of having been involved in homeland security in both the United States and Israel offered me a unique perspective that could shed light on questions of extraordinary importance.

Once we agreed on the project, we decided that the book would be a mixture of professional experiences, personal reflections, and academic scholarship. In addition, we decided that input from those "in the field" would be particularly important in introducing much needed realism to the discussion. I was extremely fortunate that individuals involved in various fields of homeland security graciously agreed to respond to questionnaires and emails. Their willingness to answer questions and contribute commentary is essential to how I view the purpose of this project. In addition, I benefited from the input of individuals presently employed by the U.S. government in the homeland security infrastructure; their willingness to meet with me and share their perspectives was predicated on assurances of anonymity. Obviously, that request will be honored. Needless to say, I am most grateful to all who took time from their busy schedules to meet with me.

This is not intended to be an abstract commentary on homeland security; rather, the project's purpose is to concretely address highly relevant issues demanding discussion, if not resolution. The scope of the book is broad, addressing legal and policy issues alike; that is critical, while recommending concrete measures intended to resolve various weaknesses and conundrums.

This project was greatly facilitated by the input—on innumerable levels—of two individuals: my research assistant, Vanessa Clayton (S.J. Quinney College of Law, University of Utah; JD 2011), and Mark Listewnik. When I (finally) agreed to write the book, I told Mark his online/on-time editorial input was essential; a book of this nature—from my perspective—demands constant interaction between writer and editor. We developed a system whereby Mark would comment both in writing and by regular phone meetings on the chapters. Obviously, we did not always agree; the reader benefits from Mark's healthy and welcome disagreements and convictions. When Vanessa agreed to work on the project, I told her my expectations were clear: honest criticism, outstanding research, and unflinching editorial comments regarding chapter drafts. To my good fortune, Vanessa hit a home run with respect to all three, and for that I am most grateful.

While Mark and Vanessa are to be credited for their significant contribution, all faults lie solely with me. Writing this book has enabled me to explore issues of deep concern to millions of people; it is my modest hope that this project will facilitate a deeper understanding of those very issues.

Amos N. Guiora

Salt Lake City

INTRODUCTION

I come to this book from four distinct perspectives: from 1997 to 2001 I served as the legal advisor to the Israel Defense Forces Home Front Command (equivalent to the U.S. Department of Homeland Security); in 2007–2008 I served as the legal advisor to a U.S. Congress-mandated task force charged with developing America's homeland security strategy; my academic scholarship has addressed various issues tangential to homeland security; I have interacted with corporate leaders and law enforcement officials, domestically and internationally, with respect to homeland security.

The breadth of relationships and contacts I have developed through these experiences undoubtedly affords me unique insight into homeland security. It also leads me to the unequivocal belief that a decade after 9/11, we have yet to sufficiently understand homeland security and threats posed. Although billions of tax dollars are yearly apportioned to homeland security, multitudes of PowerPoint presentations are prepared, and public officials make proclamations regarding homeland security on a daily basis, what is "missing in action" is clear: concise and consistent definitions.

Defining homeland security policy and strategy is essential; otherwise, both will be inherently amorphous and malleable. While fluidity may benefit decision makers, it ill-serves the public. In my testimony before a subcommittee of the U.S. Congress's Committee on Homeland Security, my recommendation to engage in a comprehensive definitional discussion was met with varying degrees of uncomfortable body language. This hesitation to clearly define key elements of homeland security is most unfortunate; it is one of my principal motivations for writing this book.

Successfully defining various aspects of homeland security—subject to a three-legged approach based on a legal, policy, and comparative analysis—requires addressing a wide range of issues. Readers may disagree with me regarding the issues I have chosen to emphasize; that is both legitimate and healthy. My hope is that this sparks additional debate. There is, after all, no "right-wrong" regarding which issues are discussed; there is, however, an obligation to analyze the issues candidly and with great scrutiny. That is my promise to the reader.

I have chosen to include the following topics, arranged by chapter:

1. What is homeland security? What and how are we seeking to protect?
2. Cost–benefit analysis of homeland security
3. Prioritizing risks/threats/dangers
4. International cooperation/intelligence sharing (the role of law enforcement)
5. Immigration and narcoterrorism—a different threat
6. The new face of terrorism—domestic threats and civil liberties
7. Terror financing—money as the "engine" that facilitates terrorism
8. Business continuity
9. Conclusion: What does the future bode?

In order to ensure that this book is not a mere abstract academic exercise, I have reached out to survey a broad array of individuals involved in homeland security. In many ways, those individuals—from the public and private sector alike—comprise an informal (sometimes anonymous) board of advisors reflecting disparate and competing interests. Though not an empirical study, answers to the questions posed were extraordinarily helpful in facilitating both the framing and understanding of the issues. Furthermore, the answers illuminate how respondents suggest prioritizing homeland security-related issues. One of the principal themes of this book is how we collectively, as a nation, can more effectively prioritize—in the context of a cost–benefit analysis—limited resources to provide policy makers with the information and tools to better secure the country.

While final responsibility of analysis and recommendations within this book rests with me, the input I received was of extraordinary assistance in facilitating greater appreciation for how issues are perceived and the degree of nuance with respect to which divergent opinions can be held regarding the same issue. Precisely because there is neither right nor wrong (unless illegal) with respect to what homeland security issues are to be addressed, it is important to be sensitive to competing perspectives. To that end, I posed the following questions to a broad range of professionals:

1. What is homeland security?
2. How do government policies (related to homeland security) impact your corporation?
3. How does your corporation assess threats/risks/dangers?
4. How does your corporation prioritize threats/risks/dangers?

5. Are you satisfied with the present level of private–public sector cooperation with respect to homeland security?
6. How do post-9/11 financial regulations impact your corporation?
7. How does your corporation facilitate business continuity (in response to either a natural disaster or terrorist attack)?
8. How does your corporation screen/vet both potential and current employees?
9. If your corporation has either international transactions or international presence (i.e., employees), how is asset protection impacted by homeland security?
10. Looking forward, what are the most important homeland security issues your corporation faces?

While the questions were slightly tweaked for law enforcement officials, the gist remained the same. The answers, which have been incorporated into the text, were revealing: they demonstrated the disparate manner in which homeland security issues are understood by individuals directly affected by and responsible for homeland security in their respective professional entities and cultures. In many ways, the range of their responses—while admirable for their candor—highlighted a significant reality: there is no universal consensus neither on what homeland security is nor on what should be the predominant considerations when analyzing homeland security, however defined. Precisely because the threats—discussed in subsequent chapters—are undeniable, defining homeland security is essential to develop both government policy and strategy for prevention, protection, and response.

There is, of course, a paradox: while decision makers can—and do—discuss homeland security endlessly, there are important limits imposed on the degree to which proposed measures can be implemented. The limits are based on law, public opinion, and enforcement ability. Declaring the desire to undertake a particular action does not guarantee the ability to implement that policy. Competing interests are inherent to any measure government seeks to impose. In the current, post-9/11 culture, homeland security is at the crux of the public debate regarding the limits of state power and the rights of the individual. Balancing the two presents extraordinary challenges to decision makers and the public alike. The controversy surrounding Arizona Senate Bill 1070 is but a mirror of the tension inherent to homeland security issues, for it manifests the "limits" discussion so essential to the debate regarding homeland security. To understand these limits, it is necessary to broadly examine issues assumed to be, directly and indirectly, homeland security.

On May 1, 2011 Osama Bin Laden was killed by Navy SEALS; Bin Laden was tracked over a number of years by U.S. intelligence agencies who discovered that he was living in Pakistan. The attack, as I commented elsewhere,* was an operational and intelligence-gathering success. However, the question going forward are the geo-political and geo-strategic implications and ramifications of the attack. The question is posed neither from a legal or moral perspective; with respect to both the 'hit' on Bin Laden was justified. Nevertheless, given uncertainty regarding Bin Laden's operational capability in May 2011—in direct contrast to a number of years ago—the larger question is whether there is a risk in turning a symbol into a martyr.

While this book's focus is neither on Bin Laden nor exclusively terrorism, Bin Laden's killing may have a direct impact on homeland security in the U.S. After all, if recent domestic terrorism discussed in this book are any indication a viable and valid argument can be propounded that the killing will serve as a 'trigger' for retributive attacks. As Israel learned in the wake of the targeted killing of the HAMAS leader, Ye'chia A'yash (January, 1996), there is extraordinary risk in such an attack. After all, killing A'yash—responsible for the deaths of over 200 Israelis—resulted both in a series of horrific suicide attacks and the stunning upset loss of incumbent Prime Minister Peres to Binyamin Netanyahu.

While killing A'yash was legally and morally justified a cost–benefit analysis raises valid questions as to its effectiveness and long-term geo-strategic impact. In killing A'yash, Israel was able to penetrate his 'inner circle' previously considered impenetrable; furthermore, the killing clearly demonstrated that even the most protected terrorists are vulnerable. The same argument can be made with respect to Bin Laden who in addition to his seemingly impenetrable inner ring was also arguably protected—whether directly or indirectly—by the Pakistani government. These issues are of the utmost importance from the perspective of national security and effective counterterrorism.

As to their relevance and applicability to homeland security, the question is whether killing Bin Laden creates, unintentionally, a new threat. After all, a terrorist organization and its disparate supporters whose leader (whether actual or symbolic) has been killed are akin to a wounded animal. It is important to recall that terrorist organizations

---

* http://jurist.law.pitt.edu/forum/2011/05/amos-guiora-targeting-bin-laden.php, last viewed May 11, 2011

'play' to a number of audiences; arguably the most important (other than the victims) are their supporters who must be convinced that the organization is vibrant, thriving and poses a danger to the identified enemy.

While this does not, obviously, mean that a domestic terrorism attack is in the offing it clearly raises its specter. Whether in the form of coordinated, sophisticated attacks on Amtrak as documents seized from Bin Laden reflect or a lone wolf attack as suggested by various experts, the Bin Laden killing unequivocally requires the Administration and law enforcement agencies to re-evaluate the state of their preparedness.

Unfortunately and inexplicably some 'experts' have surmised that the killing of Bin Laden suggests that 'the end of terrorism' is near. Nothing could be further from the truth. Rather, as someone commented in response to an op-ed I wrote, "this is only the end of the beginning."*

Therefore, as these lines are written it is too early to predict with any sense of confidence how the Bin Laden killing will impact homeland security; it is clear that enhanced vigilance must be the operational watchword. THAT has a direct impact on homeland security for the following obvious, yet complex reason: given that the Department of Homeland Security is tasked with protecting 18 critical infrastructures, the added burden of enhanced vigilance while, perhaps, force decision makers to do what is historically ignored: engage in sophisticated risk assessment predicated on complex resource allocation and prioritization with a careful eye to cost–benefit analysis.

Because this book is based on my professional experiences and academic research, writing and lecturing, the reader will note the following methodological approach: footnotes will be kept to a minimum (an exception is the input received from those who responded to the questionnaire, whether for attribution or anonymously), the writing style is more informal than formal, and comparative analyses will be heavily interspersed throughout the text. Because I am deeply skeptical of a government's ability to both articulate and implement homeland security policy, this book will be very critical of where we are. It is my hope to offer suggestions of where we could and, more importantly, should be.

---

* Private email in my records.

ABOUT THE AUTHOR

Amos Guiora is a professor of law at the S.J. Quinney College of Law, the University of Utah. Guiora, who teaches criminal procedure, international law, global perspectives on counterterrorism, and religion and terrorism, incorporates innovative scenario-based instruction to address national and international security issues and dilemmas. He is a Member of the American Bar Association's Law and National Security Advisory Committee, a research fellow at the International Institute on Counterterrorism, the Interdisciplinary Center, Herzeliya, Israel, a corresponding member of The Netherlands School of Human Rights Research, University of Utrecht School of Law, and was awarded a Senior Specialist Fulbright Fellowship for the Netherlands in 2008. Professor Guiora has published extensively in both the U.S. and Europe on issues related to national security, limits of interrogation, religion and terrorism, the limits of power and multiculturalism, and human rights. He is the author of Global Perspectives on Counterterrorism, Fundamentals of Counterterrorism, Constitutional Limits on Coercive Interrogation, and Freedom from Religion: Rights and National Security. He served for 19 years in the Israel Defense Forces as Lieutenant Colonel (retired), and held a number of senior command positions, including commander of the IDF School of Military Law and legal advisor to the Gaza Strip.

1

What Is Homeland Security?

While there is prodigious 9/11-inspired literature addressing innumerable aspects of U.S. homeland security, insufficient attention has been paid to the interplay between law and policy regarding homeland security. Each, individually, is obviously essential; however, to be effective, the two must be thoroughly enmeshed. Otherwise, policy may well violate the law, and law will be divorced from policy. In addition, examining approaches adopted by other nations facilitates developing an effective homeland security policy rooted in the law. The age-old adage "no one person has all the answers" is undoubtedly applicable to homeland security policy.

When the Bush administration established the Department of Homeland Security, it was in a "response mode"; the ashes of 9/11 were both literally and figuratively smoldering with an outraged public demanding action. President Bush's combative words directly led to a series of decisions that have, over the course of time, raised numerous questions regarding both their legality and wisdom. As these lines are written, it is unclear how President Obama will address a wide range of homeland security issues; what is clear, as will be analyzed in this book, is a continued lack of focused strategy. It is my hope to suggest strategic solutions to unsuccessful tactics that have largely characterized the two administrations.

What is clear is that the threats of yesterday have been replaced by new, ever more dangerous threats that demand our attention, both short term and long term. The threats are multifaceted and complex, forcing decision makers to prepare for previously unforeseen dilemmas and paradigms. To that end, we must ensure that homeland security measures are sufficient and viable, subject to limits imposed by the law, political considerations, and financial realities.

Herein lies a major challenge: (1) determining how we intend to prevent threats, with the understanding that not all threats can be prevented, and (2) what measures we intend to take in response to an attack or disaster, with the understanding that not all desired measures can be taken. That is, while decision makers may promise the public they are doing everything to prevent an attack and will spare no effort in responding, reality is infinitely more nuanced and complicated.

Directly addressing nuance is essential to a mature discussion regarding homeland security—the public deserves infinitely more sophisticated and robust discourse regarding issues critical to its protection and welfare than national, state, and local politicians presently offer. The 20-second sound-bite culture ill-serves the public; it also reflects a basic inability—perhaps unwillingness—to address core, complicated issues that are not black and white but instead shades of gray. A few years ago I wrote an op-ed regarding airport security; the editor's choice for the headline was "What Have We Learned?" While intended to be catchy, reality was reflected in that phrase, for one of the principal questions this book will address is just that: What have we learned since 9/11, and have we truly institutionalized lessons learned?

There have been two overwhelmingly significant homeland security events in the past 10 years: 9/11 and Katrina. In both, the U.S. government was woefully unprepared; the responses demonstrated systemic incompetence and failure. President Bush's comment to then FEMA director Michael Brown—"you're doing a heckuva job, Brownie"—distressingly symbolized a president and administration overwhelmingly failing the public.

In both instances there were significant failures in preparation and response alike. Both reflected in a failure to properly coordinate between local, state, and federal officials, and showed an inability of federal, state, and local agencies to effectively communicate and a stunning failure to determine who does what. While first responders were heroic in their responses, sacrificing themselves in the line of duty, for which they deserve the nation's never-ending gratitude; they were failed by command—political and otherwise—which had failed to foresee, much less sufficiently prepare for, such an event.

These failings were symptomatic of a fundamental failing on multiple fronts. The inability of New York City firefighters and policemen to communicate because they were literally on different frequencies is but a tragic example. The fact that the mayor of New Orleans and governor of

2

Louisiana had a mutual antipathy should never have affected the rescue efforts. However, precisely because the infrastructure was so fundamentally lacking, their personal relationship came into play.

However we define homeland security, it is incumbent that public officials recall their primary obligation is to protect the public. In order to better understand the scope of homeland security, it is necessary to engage in a definitional discussion. That will be this chapter's focus, which provides structural underpinning for the book. Because this book seeks a wide range of readers, the legal issues will be combined with significant policy discussion. As I argued before Congress, the two are inherently intertwined; understanding one requires analyzing the other. While we may disagree on how to define homeland security, I would suggest that we need to ask ourselves: Just what do we expect government to do in order to protect the public?

Contrary to those who suggest government can do it all, the reality is that limited resources must be judiciously allocated based on careful threat analysis. That means that calculated risk taking is inherent to the discussion; not every bridge, schoolhouse, airport, chemical plant, train station, nuclear facility, levee, and office building can—or should—be protected to the same level. That is impossible, and it is also, frankly, unnecessary.

THREAT

Five minutes from our home in Israel is the local elementary school our three children attended. If the school—which has an armed guard posted in its outer perimeter, buttressed by an armed, mobile patrol—were to be attacked by terrorists, it is fair to assume that the government would order the Israel Defense Forces (IDF) to aggressively engage those responsible, wherever located. The public would, by all estimates, be supportive of such a military response. By comparison, the overwhelming majority of American schools are not protected (the exception is inner-city schools, where police monitor hallways to protect against gang and drug-related violence). Friends and colleagues in the United States have commented that were their children's schools to be protected by armed guards, the children would be scared rather than necessarily feeling safer and protected. This sentiment is fair enough, since different cultures have different contexts, circumstances, approaches, and psychologies. However, this distinction goes to the core issue: Against what do we protect, and what is security, much less homeland security?

3

It is a reasonable assumption that were al-Qaeda to attack a local elementary school in an American suburb (the equivalent of where I live), parents, community leaders, and the media would lambaste the local school board, law enforcement officials, and "Washington" for failing to protect their children. The blame game ritual is well known, almost expected, as it is an inherent part of the political culture. Nearly everyone engages in this blame game—editorials and talk show pundits are predictable in their posturing, while bloggers and news stations engage in the frenzy, responding to an insatiable need for news and commentary. This is all beside the point, for "Monday morning quarterbacking" is mere noise. The fundamental question is whether a viable, perceivable threat was missed.

I begin with threat because of a belief that in discussing homeland security we must ask: What assets—human, physical, or otherwise—are we protecting, and from what are we protecting those assets? A thorough search of how different local, state, and federal agencies define homeland security results in a profound lack of clarity and absolute confusion. That definitional weakness has resulted in inconsistent government policies precisely because it is unclear what action should be taken, when it should be taken, and against whom. The result is a tactic-based prevention/response model rather than one predicated on strategic considerations. It is therefore all but impossible to determine effectiveness and efficiency, much less to gauge whether any proposed quantitative measuring instrument is genuinely reasonable or appropriate.

HOMELAND SECURITY

Herein lies a dilemma: whether to define homeland security akin to obscenity (e.g., Justice Stewart's "I know it when I see it'") or akin to terrorism (there are 109 generally accepted definitions). When confronted with this methodological question in previous scholarship, I have chosen both tracks: in a book addressing religious extremism,[†] I chose not to define religion, whereas in a book on terrorism,[‡] I chose to define terrorism. While both approaches are reasonable and justifiable, I have chosen the latter for this book because of a deep belief that wiggle room created by a lack of

---

[*] Jacobellis v. Ohio, 378 U.S. 184 (1964).

[†] Amos Guiora, Freedom from Religion: *Rights and National Security* (Oxford University Press, 2009, second edition, 2011).

[‡] Amos Guiora, Global Perspectives on Counterterrorism (Aspen Publishers, 2007, second edition, 2011).

4

definition has ill-served the public. My testimony before the U.S. Congress powerfully reinforced the absolute need to define; otherwise, government policy will never truly be policy; it will be nothing more than empty (and expensive) rhetoric. To that end, I define homeland security as:

> The protection, based on threat analysis, of domestic infrastructure and institutions whose vulnerability significantly endangers civilian life and property.

While admittedly wordy—and therefore not fitting into our 20-second sound-bite culture—the definition facilitates guidelines determining what merits protection domestically. This proposed definition will serve as the basis for issues discussed in this book; articulating and implementing rationally based guidelines is essential to developing a homeland security strategy.

By analogy, and vital to my conviction that definition is essential, in the innumerable meetings I attended as legal advisor to the Israeli Defense Force (IDF) Home Front Command (HFC), the recurring theme was: Why are we doing this (*this* being whatever protective measure was considered) and what stands behind the decision? That is, there was constant discussion as to the *why*. While the answers were not, frankly, always compelling or convincing, the question did force senior command to articulate the basis for a decision with the understanding that law and an articulated policy were to mesh.

The *why* question was not ephemeral because both the Knesset (Israeli parliament) and Supreme Court sitting as High Court of Justice—particularly the latter—could intervene in any decision taken or not taken by the HFC. The reality of external review—in accordance with checks and balances and separation of powers—significantly encouraged rational decision making. Nevertheless, owing to a fundamental lack of clarity regarding HFC mission, my staff and I routinely gave legal opinions on an extraordinary range of issues, many clearly falling outside the boundaries of the proposed definition above, precisely because insufficient time and effort had been allocated for the difficult strategic questions.

The HFC's willingness to assume responsibility for a wide range of issues directly resulted from an institutionalized failure to systematically define the term. The result was a bottomless pit predicated on the need to protect the public. While protecting the public is, indeed, the primary responsibility of government, there are limits to what government can do. The "limits" theme is essential to understanding both the articulation and implementation of homeland security.

The only manner in which those limits can be discussed cogently and coherently is by understanding that some risks or threats can neither be abated nor attended to. While this reality is perhaps harsh to some, a rational-based cost–benefit analysis of homeland security is necessary. By limiting the scope and range of homeland security, government will more effectively address prevention and response, both core issues critical to protecting the public. In adopting such an approach, the initial decision must be what is—and just as importantly, what is not—homeland security. Simply put, limiting the definition of homeland security will actually increase the level of protection.

Before answering the above questions, we need to step back and ask ourselves more strategic questions. The temptation—I have seen this both in the United States and Israel—is to "jump in" and avoid the more profound, larger, and frankly more important dilemmas. This is not akin to "shoot first, ask questions later," but rather reflects a fundamental unwillingness to engage in a strategic discussion. I have often asked myself why decision makers largely refuse to analyze homeland security questions from the perch of strategic analysis, rather preferring an action-based tactical response.

While there are a number of possible answers, I think the most convincing motivation is the rote response, "We will do whatever is necessary to save lives," as that both is good theater (back to the 20-second soundbite culture) and sends a demonstrative message to the public of government's responsibility and capability. However, as innumerable events have demonstrated, action does not suggest competence or effectiveness. While action—however defined—is compelling, it begs the larger, definitional question and merely postpones the debate we must have.

A two-star Israeli general once commented, "We avoid the difficult, strategic questions." This sort of attitude results in tactical decisions devoid of analysis of long-term ramifications and implications. The discussion is further hindered in the United States by the culture of political correctness, which limits asking the truly difficult questions—which leads us to situations like Fort Hood.

While the tragedy of Fort Hood on November 5, 2009, is indisputable, what have gone unaddressed are larger questions, which have a significant "discomfort" factor attached to them. They are, however, essential to better understanding homeland security. Major Nidal Hasan was, according to media reports, a devout Muslim. According to the U.S. Constitution, that was within his right. However, and the caveat is essential, Hasan was engaged in dialogue with an imam known to be an extremist. According to undisputed media reports, various law enforcement agencies were aware of the ongoing relationship (email exchanges) between Hasan and the imam.

While government officials could have confronted Hasan regarding this relationship in an effort to determine whether it posed a potential threat, the decision was made not to. Rather, for reasons suggestive of sensitivity to matters of faith, an easily ascertainable matter was deliberately ignored by individuals mandated with protecting the public. As I argued elsewhere,[*] while religion does not pose a danger to civil society, religious extremism clearly does. To that end, in order to protect the public (whether broader or group specific), government has the absolute obligation to act proactively in order both to ascertain the depth of the threat and, if required based on credible evidence, to take measures to minimize its possible effect.

The Obama administration's decision to deliberately obfuscate the danger posed by Islamic extremism[†] is symptomatic of political correctness. This has significant operational implications as, while it patently avoids addressing an uncomfortable subject for some, it equally does not directly and candidly address a threat. There are, as will be discussed in this book, multiple threats decision makers must currently address; Islamic extremism, with the emphasis on *extremism*, is one of those threats.

While under no circumstances is it the sole threat, it is nevertheless a major threat. To suggest otherwise, as Attorney General Holder disingenuously suggested and as reinforced in President Obama's 2010 National Security Strategy (NSS) document, is to ill-serve the American public.[‡] Removing the terms *jihad* and *Islamic extremism terrorism* from the NSS suggests turning a blind eye and subtly encouraging law enforcement officials to divert limited resources from a viable, palpable threat that has conducted significant acts of terrorism on U.S. soil, not to mention against U.S. targets overseas and against US allies.

Hasan did not—according to media reports—intend to attack the general public; rather, his intended targets were exclusively U.S. military personnel. Nevertheless, homeland security, as I have defined it, includes protecting institutions (general and specific) provided the threat sufficiently threatens life or property. Failing to make that determination required that appropriate law enforcement and intelligence community

---

[*] Amos Guiora, Freedom from Religion: Rights and National Security (Oxford University Press, 2009, second edition, 2011).

[†] http://dailyradar.com/beltwayblips/video/eric-holder-refuses-to-say-radical-islam/ (last accessed July 11, 2010).

[‡] http://www.youtube.com/watch?v=wYoVf4Aof10 (last accessed July 11, 2010).

18 Critical Infrastructure

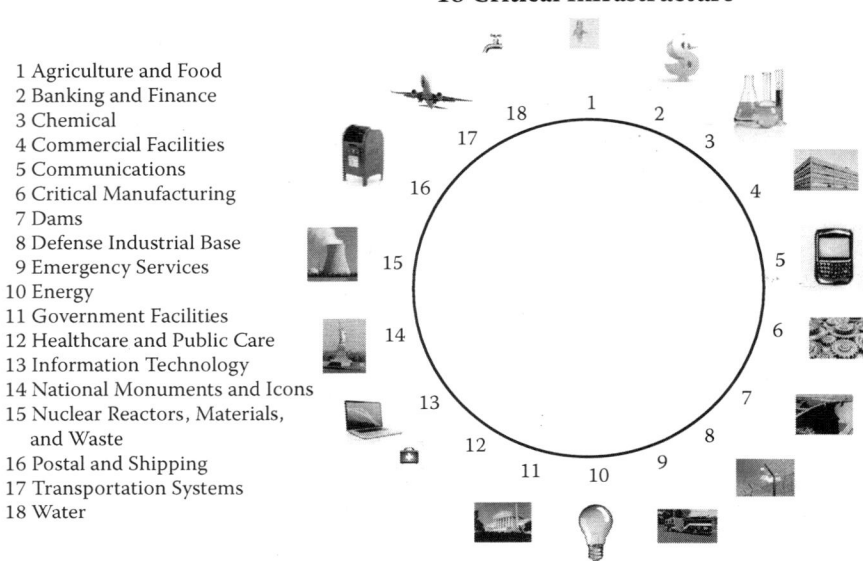

1 Agriculture and Food
2 Banking and Finance
3 Chemical
4 Commercial Facilities
5 Communications
6 Critical Manufacturing
7 Dams
8 Defense Industrial Base
9 Emergency Services
10 Energy
11 Government Facilities
12 Healthcare and Public Care
13 Information Technology
14 National Monuments and Icons
15 Nuclear Reactors, Materials,
 and Waste
16 Postal and Shipping
17 Transportation Systems
18 Water

FIGURE 1.1   The 18 critical infrastructures.

authorities either disregarded the obvious threat posed by Islamic extremism or were deterred from doing so. In either case, the system failed; if the former is true, then because of disconcerting incompetence, and if the latter, because of narrow and misbegotten political considerations.

We will return to domestic Islamic terrorism, but for now, let us turn our attention to the Department of Homeland Security (DHS).

According to the Homeland Security Presidential Directive 7 (HSPD-7), there are 18 critical infrastructures and key resources essential to the nation's security, public health and safety, economic vitality, and way of life.[*] All 18 critical infrastructures are illustrated in Figure 1.1.

I contend that these are not critical infrastructures. Rather, this is an extraordinary wish list bordering on the irresponsible: suggesting that *all* are critical infrastructures is the manifestation of "policy" devoid of any strategic consideration. It is, by analogy, what Barbara Tuchman

---

[*] http://www.dhs.gov/files/programs/gc_1189168948944.shtm (last accessed December 22, 2009).

referred to as "the march of folly."* Policy requires articulating limits; a presidential directive with 18 critical infrastructures and key resources is a nonstarter. It is a nonstarter because it mandates the Department of Homeland Security accomplish the impossible and, in an effort to do the impossible, will invariably result in missing genuine, existing threats. After all, a presidential directive is just that—a directive issued by the president of the United States to the executive branch for which officials will be held both accountable and responsible by the president and the Congress. In the context of separation of powers and checks and balances, Congress demands accountability; that is the essence of congressional oversight which requires officials in the executive branch testify before congressional committees. The House has a House Committee on Homeland Security and six subcommittees, while the Senate has the Senate Committee on Homeland Security and Government Affairs with seven subcommittees.†

Throughout this book I will refer to the 18 critical infrastructures, but it is essential that the reader understand that the mission (actually, missions) mandated to DHS is beyond the pale in terms of reasonableness. That unreasonable·expectation naturally translates into an enormous budget—nearly $42.8 billion for 2009.‡ It is one of my principal contentions that the enormous budget allocated to DHS reflects a fundamental failure of Congress and the executive branch alike, for it creates a "mission impossible," whereby goals and objectives are inherently unrealistic as, in large part, prioritization is disconcertingly divorced from the process.

In February 2008 I was asked by the Committee on Homeland Security's staff to prepare questions in anticipation of Secretary Michael Chertoff's testimony. Chertoff had been invited by the committee—in its oversight capacity—to address budgetary matters and to provide a status report on a wide range of issues. The questions I prepared were focused on the requirement to develop a matrix-based cost–benefit analysis of homeland security that would enable articulation and demonstration of effectiveness. In addition, akin to my testimony before the committee, the questions focused on the need to define terms relevant to homeland security. The theme of the questions was twofold: to encourage development

---

* http://www.amazon.com/March-Folly-Troy-Vietnam/dp/0345308239 (last accessed July 11, 2010).

† http://www.contactingthecongress.org/cgi-bin/committee_list.cgi?site=ctc (last accessed October 24, 2010).

‡ http://www.whitehouse.gov/omb/rewrite/budget/fy2009/homeland.html (last accessed July 11, 2010); http://www.reuters.com/article/idUSN20448819 (last accessed July 11, 2010).

of a quantifiable matrix relevant to the DHS mandate and to define terms bandied about.

Attending Chertoff's testimony left me with a bad taste in my mouth reinforced when I read his book, *Homeland Security: Assessing the First Five Years.** Both his congressional testimony and book reflect a belief (at least for public consumption) that homeland security is all encompassing and the department is capable of performing all missions without exception. Sitting in the audience—knowing what questions would be posed to Chertoff—I was struck by the profound disconnect between his testimony and operational reality. The disconnect was premised on my belief that homeland security—in accordance with my proposed definition—must be specific threat predicated rather than based on an overencumbered, impossible to define, much less achieve, "catchall." While Chertoff was articulate, prepared, and impressive, his testimony was ultimately unrealistic; the underlying premise as mandated in the presidential directive was not within the range of goals that even determined and dedicated public officials—regardless of their efforts—could reasonably achieve.[†]

While Chertoff could highlight department achievements and committee members could either criticize or laud him (depending on their political affiliation), none addressed the core issue: What is homeland security? Lacking in the give-and-take between Chertoff and the committee was a sophisticated discussion; tactics dominated, with strategic considerations not addressed. For me, it had an all too familiar ring: in the 4½ years I served as the legal advisor to the IDF HFC, the same fundamental flaw was omnipresent. The need to create and maintain an oversized, largely unmanageable umbrella agency encompassing distinct, if not competing, organizational cultures was the operational reality.

It is important to emphasize that Fort Hood was not the exclusive responsibility of the Department of Homeland Security; any number of other federal agencies (including the military) evidently missed warning signs. While internal and external investigations alike will determine whether the tragedy could have been prevented, what is important for our purposes is why *proactive* measures were not taken.[‡] That question

---

[*] Michael Chertoff, Homeland Security: Assessing the First Five Years (University of Pennsylvania Press, 2009).

[†] Difficulty to achieve the goals is heightened by the extraordinarily low morale of DHS staff; http://www.centerforinvestigativereporting.org/blogpost/20090115homelandsecurityusatheouttakesparti (last accessed July 11, 2010).

[‡] http://www.msnbc.msn.com/id/33753461/ns/us_news-tragedy_at_fort_hood/ (last accessed December 24, 2009).

is asked in the context of the two critical aspects of homeland security: prevention and response in the face of a viable threat.

While I clearly advocate adopting a limited approach to homeland security, I simultaneously argue that threats deemed viable must be addressed aggressively and consistently even if certain segments of the population will be offended. There is a fine line between constitutional violations, which I vociferously reject, and aggressive law enforcement within constitutional boundaries.* As will be discussed in subsequent chapters, many of the Bush administration's post-9/11 decisions clearly violated otherwise constitutional protections; worse than this, many of those actions were also ineffective. Therefore, the theme of limits must be understood as cojoined with lawful, aggressive measures. The fear of offending, predicated on pervasive political correctness, is a major stumbling block to effective homeland security tactics and strategy alike. To that end, as will be explored in subsequent chapters, homeland security need not be meek or submissive; however, it must be limited and based on strategic considerations.

Therefore, limits is a recurring theme in this book; if we fail to understand and address this core concept, those responsible for homeland security will constantly and continuously be incapable of answering the simple question: What is homeland security? In numerous conversations with law enforcement officials, my question "What is homeland security?" is generally met with a shrug of the shoulders and a brief response of "good question." More often than not, that was the response of my colleagues at the HFC. The following questions are but a mere sample of *possible* issues that could be deemed relevant to homeland security; the ultimate question is: Which of them should be, and are addressable, and at what cost? They are intended to facilitate the reader's appreciation for homeland security dilemmas; they also suggest the scope of contemporary homeland security issues:

1. Should all passengers departing from U.S. airports be equally subject to TSA screening procedures and should standard screening include full-body scanners?
2. Should the FBI have access to an individual's public library records?

---

* For a fuller discussion of the limits issue, see Amos Guiora, Constitutional Limits on Coercive Interrogation (Oxford: Oxford University Press, 2008).

3. Should armed guards protect the nation's schools?
4. Should individuals attending sporting events be subject to pat downs and frisks?
5. Should the size of water bottles carried on airplanes be limited?
6. Should the amount of financial wire transfers be limited and subject to reporting requirements?
7. Should all individuals entering corporate headquarters/office buildings be required to register their names?
8. Should law enforcement monitor emails of private citizens, and if yes, subject to what criteria and limits?
9. Should law enforcement monitor/conduct surveillance of houses of worship?
10. Should immigration policy accommodate illegal immigrants or should a no tolerance policy be implemented?
11. Should all chemical plants be protected, and if yes, should the protection be equal?
12. Should pandemic planning preparedness and response be under the auspices of Homeland Security?

Since 9/11, these dilemmas have become an inherent part of American life; they reflect how homeland security questions have come to dominate the daily landscape. Those who fly regularly know the drill: take off your shoes, unfasten your belt, take out your laptop, remove your jacket, and make sure to have packed toothpaste of a certain size in a plastic bag. While the Transportation Safety Authority (TSA) is mockingly referred to as "thousands standing around," the humor is not misplaced. On more than one occasion I have wondered whether the long lines at security checkpoints do not invite acts of terrorism. As passengers stand shoeless, more often than not with their backs to the airport's all but unprotected front doors, there is a certain incredulousness on my part. I ask myself: Are we so convinced that there will be an attack on a plane that we make ourselves vulnerable while waiting to board that plane?

Similarly, America's shopping malls are overwhelmingly unprotected largely, I suggest, because of fear of inconveniencing shoppers. Different cultures, different paradigms: friends in Israel will not eat at a restaurant if there is no guard; friends in America recoil at the thought of having their bag searched before dining out. Similarly, friends in Israel ensure their children are aware of threats and dangers; friends in America wonder whether to share news of terrorist attacks with their children.

An important aspect of homeland security is educating the public; to that end, government must make better efforts to inform regarding various risks, ranging from terrorism to public health to levees. To do otherwise is irresponsible and leads to unnecessary panic when decisions—otherwise seemingly random and arbitrary—are announced, sometimes devoid of previous context.

As citizens of New Orleans were warned to leave in anticipation of Katrina, the pictures on TV were surreal: those with means left in an orderly manner; those without means either "stuck it out" or were left in an extraordinarily haphazard manner recalling pictures of South Vietnamese clinging onto American helicopters on the rooftop of the American embassy in Saigon begging to flee. Akin to watching a Greek tragedy where the end is known, Katrina devastated New Orleans. While Mayor Ray Nagin vastly exaggerated the number of casualties, the essence of his misinformation, while remarkably unprofessional, cannot be dismissed. The mayor's inability to provide correct information reflected a breakdown of systems expected to operate in times of crisis.

It is critical to understand that *effective* (the term will be discussed in full in subsequent chapters) homeland security demands an integrative approach. What I refer to as prevention and response extends to a wide range of disciplines, fundamental among them informing the public of an event, instructing them how to respond, and if need be, directing them where to go. I emphasize this because the client in homeland security is the civilian population. While the goal is to protect, the question is at what cost to liberty and freedom. This is not an amorphous question, for one of the dilemmas in homeland security is to what extent can otherwise guaranteed rights be limited in order to protect the population, either in the face of a future event or in response to an occurring event. The need to directly address the civilian population, therefore, is an essential aspect of homeland security.

Unlike war, which occurs in far-off battlefields, homeland security happens—literally—in our collective and individual backyards. The physical proximity between a threat or actual event (terrorism or natural disaster) and the civilian population is critical in understanding the visceral reality of homeland security. Precisely because domestic terrorism is just that—terrorism that occurs domestically—the requirement to directly communicate with and educate the public is fundamentally different than the war paradigm. In developing homeland security strategy and policy, decision makers must place the public "front and center," with respect to both protection/prevention and explanation/education. To that end, the HFC had a department whose sole responsibility was to prepare and educate the civilian population in the event of an unconventional attack. By contrast, when I participated in a simulation exercise with physicians in Cleveland, Ohio, informing the public in real time was, at best, an afterthought.*

In direct contrast, in a complicated, multifaceted, interdisciplinary bioterrorism simulation exercise that I developed and conducted with local, state, national, and international officials, the roles and interaction of media and press officers were carefully examined and much discussed.† There was a clear understanding, among the participants, that communicating with the general public was an integral—if not essential—aspect of homeland security (this in direct contrast to the approach suggested in the other simulation). In this vein, it is increasingly clear that nontraditional media, in particular social networks (e.g., Twitter, Facebook, texting), facilitate the spreading of information (correct or otherwise) like wildfire. This requires proactive information dissemination by decision makers demonstrating extraordinary nimbleness and flexibility that has not been, historically, the cornerstone of government media efforts.

In addition, the multifaceted exercise emphasized the following:

1. Providing security to the nation's homeland requires government officials develop a sophisticated understanding of different population groups (e.g., religious and ethnic sensitivities, matriarchal/patriarchal nature, socioeconomic status, education levels).
2. Language proficiency is needed for non-English-speaking populations.

---

* A real-life example of this was the federal government's response to Hurricane Katrina. The failures led to reform with the Post-Katrina Emergency Reform Act; more information available at http://www.fema.gov/about/history.shtm. While our response to disaster—whether natural or acts of terror—has hopefully improved, other countries have better prepared their populations.
† Simulation available at http://www.youtube.com/watch?v=kv2Oag7UmXA (last accessed October 24, 2010).

3. The right to impose, and ability to enforce, quarantines (and other impositions on otherwise guaranteed constitutional protections) on the civilian population is integral to homeland security.
4. Private-public sector cooperation is essential.
5. International cooperation is essential, though it presents major challenges.
6. Local-state-federal cooperation is essential, though major legal and policy challenges exist.
7. Determining threat viability is essential, particularly given limited state resources.
8. Determining—in advance—temporary living arrangements for significant numbers of the civilian population requires informing the public regarding either pick-up locations (for purposes of public transportation) or alternative travel routes; this effort will be enormously facilitated if prerecorded announcements are prepared in different/community-relevant languages.
9. Homeland security is distinct from traditional law enforcement and requires additional, alternative training.
10. Cost–benefit analysis is essential for determining what possible attacks agencies must prepare for.

This 10-point take away from the bioterrorism simulation is relevant to terrorist attacks and natural disasters alike. In the spirit of the narrow definition I propose for homeland security, focusing on these will enable decision makers to more effectively and efficiently serve and protect the public. Homeland security, as we will see in the chapters to come, cannot reflect a laundry list of issues; it must be focused and narrow, cognizant of budget and resource realities. It is essential that the reader understand that many officials involved in homeland security also have day jobs; the local police chief, who in a bioterrorism attack has certain responsibilities (e.g., imposing/enforcing quarantine, maintaining order), is also charged with public safety on a daily basis. The same holds true for a multitude of officials charged with *implementation* of homeland security; unlike those who set policy, those "on the ground" are sufficiently distracted, if not overburdened, with their primary responsibilities (e.g., the police chief). To that end, I would suggest

15

that the 18 critical infrastructures/key resources outlined by/tasked to the Department of Homeland Security represent a nonstarter. How we address that, and how we determine what homeland security really is, will be our focus as we go forward in the chapters to come.

2
Cost–Benefit Analysis

We spend an enormous amount of money on homeland security—nearly $42.8 billion in 2009—and we also spend significant time on homeland security. One of this book's primary aims is to candidly address how costs *and* benefits of homeland security are determined. The use of the word *and* is essential to the discussion, for this is not a zero-sum game. Costs do not guarantee or imply benefits. Similarly, not all benefits necessarily derive from either cost or spending, and if costs are attached, they should not be unlimited.

The terrorist attacks of September 11, 2001, left Americans feeling more vulnerable than ever. By example, as often noted by media and observers alike for weeks thereafter, people walking in New York City or Washington, D.C., would glance nervously skyward in understandable reaction to noises from above. Even in seemingly safe communities across the heartland, people thought twice before going to shopping malls or restaurants.

Despite the passage of time (over nine years), numerous state and national legislative efforts, billions of dollars, and an inordinate amount of time and effort, failures that led to the attacks are as prevalent as they were in 2001. One of the primary causes of this failure is an inability or unwillingness to competently and consistently define terms essential to homeland security, including costs, benefits, effectiveness, and accountability. Failure to define terms prevents development of workable homeland security models, institutionalized measures that would facilitate implementation of strategic preparation and response measures, rather than inconsistent tactical responses.

FY 2010 Budget Overview

	FY 2008 Revised Enacted[a]	FY 2009 Enacted[b]	FY 2010 Present Budget	FY 2010 ± FY 2009
	$000	$000	$000	$000
Net discretionary	$35,065,701	$40,056,930	$42,713,922	$2,656,992
Discretionary fees	2,923,150	3,166,019	3,072,030	(93,989)
Less rescission of prior year carryover[c]	*(124,985)*	*(61,373)*	—	*61,373*
Gross discretionary	**37,863,867**	**43,161,576**	**45,785,952**	**2,624,376**
Mandatory, fee, trust funds	9,465,797	9,320,643	9,329,275	8,632
Total budget authority	47,329,664	52,482,219	55,115,227	2,633,008
Supplemental[d]	15,129,607	2,967,000	—	(2,967,000)

[a] FY 2008 revised enacted reflects net reprogramming/transfer adjustments for CBP ($2.6 million), TSA (–$10.5 million), USSS ($34.0 million), NPPD (–$5.6 million), OHA ($1.9 million), FEMA (–$23.0 million), USCIS ($282.167 million), FLETC ($5.636 million), and FEMA—DRF to OIG ($16 million). It reflects technical adjustments to revise fee estimates for TSA Aviation Security—General Aviation Fee ($0.050 million), TSA Aviation Security—Passenger and Aviation Security Infrastructure Fee ($96.025 million), TSA Transportation Threat Assessment and Credentialing—Registered Traveler (–$31.601 million), TSA Transportation Threat Assessment and Credentialing—Transportation Worker Identification Credentials ($37.9 million), TSA Transportation Threat Assessment and Credentialing—HAZMAT (–$1.0 million), TSA Transportation Threat Assessment and Credentialing—Alien Flight School ($1.0 million), and FEMA—Radiological Emergency Preparedness ($–0.492 million). Pursuant to P.L. 110-161, it reflects a scorekeeping adjustment for rescissions of prior year unobligated balances from USCG AC&I (–$137.264 million) and a rescission of current year appropriations for USM (–$5.0 million).

[b] FY 2009 enacted reflects technical adjustments to revise fee estimates for TSA—Transportation Threat and Credentialing—Registered Traveler (–$10.0 million), TSA—Transportation Threat and Credentialing—Transportation Worker Identification Credentials ($22.7 million), TSA—Transportation Threat and Credentialing—HAZMAT (–$3.0 million), and TSA—Transportation Threat and Credentialing—Alien Flight School ($1.0 million). It reflects USCG realignment of operating expenses funding and pursuant to P.L. 110-53 reflects TSA realignment of funds for 9/11 Commission Act implementation ($3.675 million, aviation security; 13.825 million, surface; $2.5 million, support). It reflects a scorekeeping adjustment for a rescission of prior year unobligated balances from USCG AC&I (–$20.0 million).

FY 2010 Budget Overview (*Continued*)

c Pursuant to P.L. 110-161, it reflects rescission of prior year unobligated balances: FY 2008—Counterterrorism Fund (–$8.480 million), TSA (–$4.5 million), Analysis and Operations (–$8.7 million), FEMA—Disaster Relief Fund (–$20.0 million), USCG—Operating Expenses (–$9.584 million), CBP (–$2.003 million), USCIS (–$0.672 million), FEMA (–$2.919 million), ICE (–$5.137 million), FLETC (–$0.334 million), OSEM (–$4.211 million), USM (–$0.444 million), CFO (–$0.380 million), CIO (–$0.493 million), DNDO (–$0.368 million), OHA (–$0.045 million), OIG (–$0.032 million), NPPD (–$1.995 million), and S&T (–$0.217 million). Pursuant to P.L. 110-161, it reflects rescission of start-up balances: FY 2008—CBP (–$25.621 million), FEMA (–$14.257 million), Departmental Operations ($12.084 million), and Working Capital Fund (–$2.509 million). Pursuant to P.L. 110-329, it reflects rescission of prior year unobligated balances: FY 2009—Analysis and Operations (–$21.373 million), TSA (–$31.0 million), and FEMA—Cerro Grande (–$9.0 million).

d In order to obtain comparable figures, total budget authority excludes:
- FY 2008 supplemental funding pursuant to P.L. 110-161: CBP ($1.531 billion), ICE ($526.9 million), USCG ($166.1 million), NPPD ($275.0 million), FEMA ($3.030 billion), USCIS ($80.0 million), and FLETC ($21.0 million)
- FY 2008 supplemental funding pursuant to P.L. 110-252: USCG ($222.607 million) and FEMA ($897.0 million)
- FY 2008 supplemental funding pursuant to P.L. 110-329: OIG ($8.0 million), USCG ($300.0 million), and FEMA ($8.072 billion)
- FY 2009 supplemental funding pursuant to P.L 110-252: USCG ($112 million)
- FY 2009 supplemental funding pursuant to P.L. 111-5: USM ($200 million), CBP ($680 million), ICE ($20 million), TSA ($1.0 billion), USCG ($240 million), FEMA ($610 million), and OIG ($5 million)
- FY 2009 supplemental funding pursuant to P.L. 111-8: USSS ($100 million)

Effective homeland security significantly impacts terrorists' infrastructure essential to perpetrating attacks. To that end, effective homeland security prevents some—but not all—attacks by either preventing a particular, planned attack from going forth or postponing or impacting plans for future attacks. For the state's effort to be truly effective, measures must reflect minimizing collateral damage, exercising fiscal responsibility, and preserving civil liberties. Security analysts and talking heads are wont to articulate effectiveness as requiring foolproof safeguards suggesting that preventing all acts of terrorism is an operational possibility. It is not.

Perhaps good theater, but in practical terms, it is but an unattainable wish. For the public to demand "no terrorist attacks" is to demand of government inherently unrealistic goals and to impose impossible objectives. However, a successful terrorist attack does not mean existing homeland security measures are ineffective. The inverse is also true: the absence of terrorist attacks does not necessarily indicate existing counterterrorism measures are effective.

Homeland security must be conducted while balancing competing interests, including determining what security measures society is willing to tolerate in the context of convenience/inconvenience, what financial costs the public is willing to absorb *not only* in the immediate aftermath of an attack, and the degree to which the infringement of civil liberties will be accepted. Balancing legitimate national security considerations with equally legitimate individual rights is the most significant issue faced by liberal democratic nations developing a viable, effective, and legal homeland security strategy. Without a balance between these two tensions, democratic societies lose the very ethos for which they fight. As Benjamin Franklin once said, "Those who would give up essential liberty, to purchase a little temporary safety, deserve neither liberty nor safety."*

Financial constraints necessarily limit measures a nation may take to protect innocent civilians, critical infrastructure, and personal property. Prioritization facilitates determining resource allocation in determining *how and when* the three can be most effectively protected. They also impose the requirement to prioritize resources and threats. To that end, given the reality of limited resources, decision makers must deliberately determine appropriate counterterrorism measures. With limited resources, government must pick and choose which measures will most effectively combat short-term and long-term threats once they have been *carefully* defined.

Caution must not be thrown to the wind, as unwarranted excess can loom just around the corner; "round up the usual suspects" must never be a policy, regardless of the threat real or imagined. Although the need for fiscal responsibility lessens as a threat becomes more imminent—prevent the threat at all costs becomes the overwhelming mentality, with tragic results inevitable. Deliberate planning based on assessment of viable threats indicative of potential attacks will facilitate judicious use of financial resources even in the face of imminent threats. This would thereby free resources for countering long-range and uncertain threats. Thus, financial responsibility needs to be considered not only in light of the threat level of a particular threat, but also in light of an overarching homeland security strategy.

---

* Benjamin Franklin, An Historical Review of the Constitution and Government of Pennsylvania, 1759.

HOMELAND SECURITY COST–BENEFITS

What is a benefit? The easy answer is that it is something an individual or society gains. Examples are abundant, ranging from financial to social to intellectual; they also include, in the context of homeland security, personal and physical safety. The question is: At what cost?

By quick example: On January 4, 2009, thousands of Newark Airport passengers were rescreened because one individual (not a traveler) entered the "sterile zone" unchecked. The resulting lockdown significantly enhanced danger to those very people.* The bungled and incompetent effort to find the individual magnified risks for passengers who had already been screened but had to be reprocessed, thereby creating both a security and a safety hazard, not to mention a huge inconvenience to the passengers, airport, and airline staff.

TSA officials (responsible for the original "slipping through") could not immediately locate the individual because the much-advertised monitors and cameras were malfunctioning. Restricting the access of individuals to a place from which they are rightly barred is Security 101. If the individual was not screened, then he should have been caught when seeking to enter the screened zone security. The original security failure was compounded by potentially endangering thousands of passengers now required to undergo rechecking under extraordinarily crowded conditions that—were the initial security violation a trap—could have directly contributed to an attack with tragic consequences. Basically, the response reflected a system that, once triggered, went into an automatic response mode. This response was devoid of both nuance and deliberate decision making. There was no analysis of any potential immediate and subsequent secondary threat (prior to such an incident happening). Meaning, if part 1 was a man entering the restricted area, intending to serve as a diversion to a larger, much more secondary breach or attack, we could have witnessed our next 9/11.

In analyzing this scenario—which fortunately ended with no harm incurred—from a cost–benefit perspective, the following questions are relevant for a sophisticated lessons learned process:

---

* http://wcbstv.com/local/newark.airport.continental.2.1407062.html (last accessed January 6, 2010).

1. In the case of a security breach, what is TSA standard operating procedure (SOP)?
2. Has the SOP (in case of breach) been examined from the perspective of the immediate (capturing the perpetrator) and the actions taken subsequently (endangering a significant population group)?
3. Does SOP include mitigation efforts intended to minimize the costs to thousands of already screened passengers?
4. What command and control measures had been previously implemented to determine the cost–benefit of the response?
5. Did SOP allow for discretion of on-site command, or does the system go into an automatic mode?
6. Had SOP been subjected to a trial-and-error process?
7. Was the SOP subject to a cost–benefit analysis?
8. How were risks and threats assessed when the SOP was developed, authorized, and implemented?
9. How was secondary risk (to travelers previously screened) defined?
10. What is SOP for determining monitors are fully functioning?

Below is another example.

In 2003 I was sent by the IDF to the United States to meet with senior military commanders. While attempting to board a flight at Ronald Reagan Washington National Airport (there used to be additional random screening at the gate that has recently been reimplemented at some airports) I was witness to the following scene: Seeking to board the plane in front of me was an elderly nun who was asked to remove her shoes. When she respectfully told the TSA official that she is unable to do so because she has a wooden leg (with a shoe attached to her full-length wooden leg) the response was to cite a regulation that mandated all shoes must be removed. When she indicated she does not have a shoe, the response was: "No shoe removal, no plane boarded." While the nun explained that taking off her wooden leg requires (1) unscrewing the leg and (2) raising her habit, I interceded and suggested to the TSA official that perhaps discretion and human dignity (polite term for common sense)

22

should rule the day. When he insisted on removing her shoe (and raising her habit) I took it upon myself to shield her from the public. A check of the leg/shoe confirmed my suspicion that she was not a terrorist. When he handed her leg back to her, the nun said, "Sir, I can't screw it back myself. You have to do that because I am physically incapable of doing so."

All travelers have been witness to similar—if not worse—incidents; some readers, obviously, are the victims of airport security that suffers from a frightening lack of sophisticated strategic thinking, much less basic common sense. In the classroom, my students are urged to view law and policy through a theory called "multiple audiences." This theory suggests that every event must be analyzed through the lens of different, perhaps competing, interests and distinct individual perspectives. The student is required to examine a factual situation predicated on concentric circles, distinguishing between those parties or individuals most directly affected by the event, moving out to those least affected by the event.

In the first example, in the incident in New Jersey, the most directly affected were the thousands of passengers rescreened and TSA officials trying frantically (according to some reports) to identify the individual who entered the screened passenger zone. The second most affected were the TSA screeners now facing a clearly unfamiliar (seemingly unsimulated and untrained for) dilemma requiring them to simultaneously rescreen thousands of passengers who were also confronted with an unexpected situation. The next group of affected individuals were those who had just arrived at the airport (some aware, others not, of the drama unfolding), and the least affected were those who had later flights.

In the second example, the one I witnessed at Reagan Airport, the most directly affected individual was the nun, followed by the regulation-abiding TSA official, and the third group were witnesses to a humiliating and embarrassing scene that—one assumes—raised concern in their minds regarding homeland security measures and policies (Figure 2.1).

While my flight (with the nun safely nestled in her seat) departed as scheduled (unlike the first example), lessons to be drawn from this event are no less important than the Newark Airport "meltdown." In both cases there was a shocking lack of discretion and common sense that resulted in costs to all categories, regardless of their proximity to the actual event. In each paradigm there are multiple costs: time to delayed

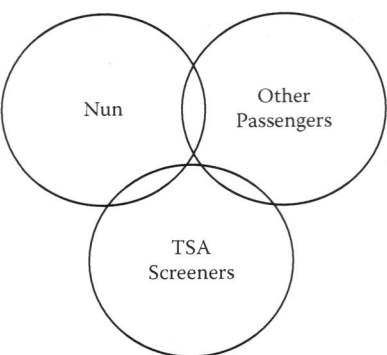

FIGURE 2.1 Overlapping circles of those affected in the Reagan Airport example.

travelers (first scenario), which has both direct costs (i.e., missed business meetings, appointments, job interviews, and flight connections) and indirect costs (the residual effects of missed connections, including possibly holding connecting flights at multiple locations, thereby directly affecting travelers not at Newark); social costs to both the nun and witnesses (second scenario); and safety/security costs inevitably arising from the rechecking, which directly impacts TSA's ability to respond to the needs and safety of passengers not yet screened but also now standing in endlessly long lines with the inevitable irritation and concern arising from such situations. Each of these costs is high; while how high depends on individual circumstances and conditions, the costs to all affected groups are high.

As to the benefits, the following questions focus the discussion:

1. Did TSA officials train (and retrain) for the first scenario?
2. What were the stated objectives (of TSA) in both scenarios?
3. How was successful resolution in the first scenario defined?
4. Were costs factored into the benefits definition in both scenarios?
5. Were either costs or benefits based on previous experiences or simulations incorporating all four, distinct population groups?

6. Was the input of "on the ground" TSA officials solicited in developing a response model (on the premise a response model had been developed)?
7. Are residual costs incorporated into the benefits model?
8. What costs are deemed reasonable/unreasonable?
9. From TSA's perspective, are defined benefits an absolute, or are shades of gray deemed operationally acceptable?
10. What will be the operational response (both strategically and tactically) in the event of a reoccurrence?

With respect to effectiveness, the politically correct spin—"We learn from mistakes and seek to improve in our efforts to protect the American public"—might satisfy some. I would suggest that camera systems that do not work and TSA officials incapable of discretion illustrate minimal effectiveness, if any at all.

The effectiveness discussion must be understood in the threat context: What threat (perceived or actual) is a particular policy or action intended to address or mollify? Herein lies the fundamental issue relevant to the cost–benefit/effectiveness question: Is there a threat that justifies implementation of particular, specific measures that limit otherwise innocent individuals? After all, homeland security must balance competing interests and concerns, but to be effective it must do so in a manner that reflects caution and reasoning. Simply put, panic responses suggest "flailing," which significantly raises the cost mirrored by a dramatic lowering—if not negating—of any possible benefit.

In Chapter 1, I suggested preserving and protecting the 18 infrastructures DHS has defined as critical reflects an unattainable goal. As the essence of cost–benefit analysis is measuring what you can achieve at what cost, the question is: What is homeland security truly seeking to achieve? If we weigh costs as identified above against DHS missions (18), then it is highly unlikely that goals can be met. The danger, however, is not insignificant; in an effort to achieve the unachievable, government will inevitably impose significant costs with minimal benefits. That is the essence of ineffective homeland security. The end game is connecting responsible spending effectively to counter real threats to protect what can realistically be protected.

Case in point: In response to the attempted Christmas Day bombing of Northwest Airlines Flight 253 at Detroit Metropolitan Airport, American officials decided, in part, to increase the number of body scanners and to subject an increased number of passengers to the scanners. While the scanners may seem to prevent passengers carrying dangerous weapons from boarding planes, the question is at what cost. That is, not only are the scanners expensive,* but they are invasive with respect to individual privacy. While TSA officials stressed that those monitoring the body scanners do not see the face of the scanned individual, thereby preventing them from connecting face to body, reality need not be gainsaid: someone is seeing the passenger's body. While some might argue that is a justified cost under present circumstances, I suggest something profoundly different.

The Nigerian terrorist sent, allegedly by al-Qaeda operatives in Yemen, to kill his fellow passengers should never have been allowed to board the flight. Body scanners—regardless of their cost—were not necessary to determine that Umar Farouk Abdulmutallab, who boarded the flight at Amsterdam Schiphol Airport after flying in from Lagos, Nigeria (Abdulmutallab transferred planes at Amsterdam), simply should not have been permitted to do so (Figure 2.2).

Abdulmutallab:

1. Purchased a one-way ticket
2. Purchased the ticket with cash
3. Had no luggage
4. Had no overcoat
5. Possibly had no passport

In other words, who needs body scanners when common sense and basic, elementary discretion should have absolutely prevented the attack. That is, maximum benefit at low cost with maximum effectiveness. The traditional American impulse to engage in overdoing and overreacting unfortunately facilitates the rush to purchase and implement additional body scanners. When President Obama announced a series of measures on January 7, 2010, I, like many, had a sense of déjà vu.[†] It was a condensed

---

* Body scanners cost between $130,000 and $170,000 to install. http://www.mercurynews.com/top-stories/ci_15279748?nclick_check=1 (last accessed October 24, 2010).
[†] http://i.cdn.turner.com/cnn/2010/images/01/07/potus.directive.corrective.actions.pdf (last accessed January 8, 2010).

Alleged Bomber's Route

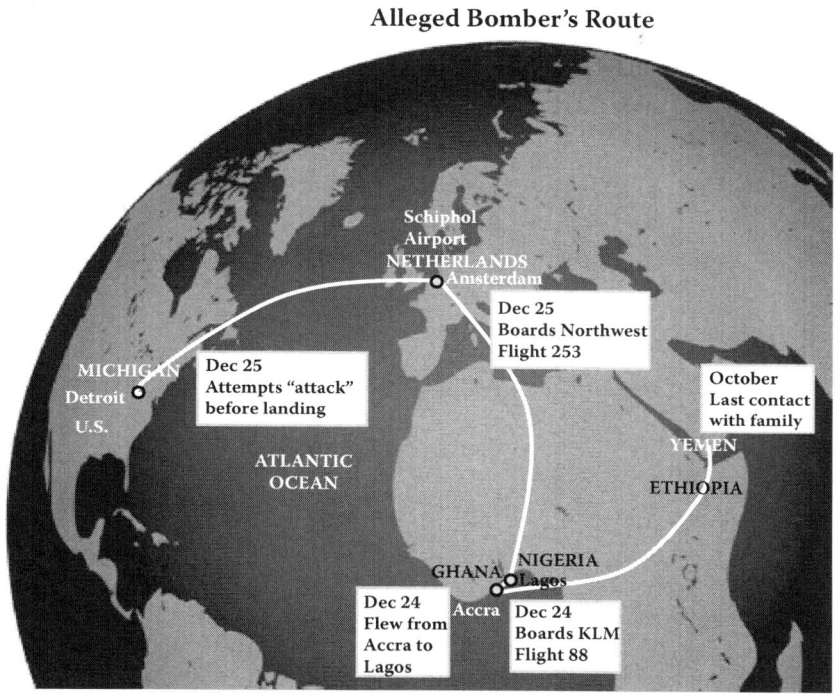

FIGURE 2.2 Umar Farouk Abdulmutallab's timeline and route from Yemen to Detroit.

form of the 9/11 Commission Report.[*] Worse than that, by emphasizing the international threats posed by al-Qaeda (which clearly exist), American officials ignored viable threats to homeland security posed by al-Qaeda.[†] That is, focusing on the operational capability of al-Qaeda to American forces in Iraq, Afghanistan, Pakistan, and Yemen should not suggest the organization does not have the ability to attack Americans in the United States.

This, naturally, brings us to a discussion of threat assessment; otherwise, the cost–benefit discussion is nothing more than an intellectual exercise. Threat assessment is an extraordinarily complex and layered process, for it involves piecing together multiple pieces of information

---

[*] http://www.9-11commission.gov/report/911Report.pdf (last accessed January 8, 2010).
[†] http://www.washingtonpost.com/wp-dyn/content/article/2010/01/07/AR2010010702310. html?hpid=topnews (last accessed January 8, 2010).

FIGURE 2.3 Jigsaw puzzle, scattered.

from various, independent sources (Figure 2.3). It is akin to a jigsaw puzzle, albeit with far greater stakes and with far less concrete information. Nevertheless, the analogy is instructive.

In order to build the puzzle (see Figure 2.4), intelligence analysts receiving information have to determine its reliability, viability, validity, and credibility; doing so requires a sophisticated understanding of the information's source. In other words, if the source is a person (referred to as human intelligence), answers to the following questions guide decision makers in determining whether the information received is "actionable:"

1. Does the source have a personal grudge against the individual he has named?
2. How is the source's reliability determined?
3. What are the criteria for recruiting and using a source?
4. How many different sources are necessary for decision makers to determine threat level?

FIGURE 2.4 Jigsaw puzzle, completed.

Test Prong	Definition/Use
Reliable	Past experiences show the source to be a dependable provider of correct information. Requires discerning whether the information is useful and accurate. Demands analysis by the case officer whether the source has a personal agenda/grudge with respect to the person identified/targeted.
Viable	Is it possible that an attack could occur in accordance with the source's information? That is, the information provided by the source indicates a terrorist attack that could take place within the realm of the possible and feasible.
Relevant	The information has bearing on upcoming events. Consider both the timeliness of the information and whether it is time sensitive, imposing immediate counterterrorism measures.
Corroborated	Another source (who meets the reliability test above) confirms the information in whole or part.

While determining the danger posed by a specific threat is an inexact science, concrete decisions regarding how to respond to the threat—if at all—are the essences of sophisticated homeland security. The relationship between cost–benefit analysis and threat levels is absolute; if threat determination is predicated solely on intuition, then the chances for overreaction all but guarantee that costs increase and benefits decrease. Conversely, if threat determination is predicated on sophisticated analysis of actionable intelligence, then chances for overreaction (and its attendant costs) are significantly minimized. While homeland security obviously implies costs (financial, legal, and social), the challenge facing decision makers is minimizing costs while maximizing benefits in an effort to protect the public.

The return on investment formula:

$$ROI = \frac{\left(\text{Gain from Investment} - \text{Cost of Investment}\right)}{\text{Cost of Investment}}$$

By way of example: In the immediate aftermath of the attempted Christmas Day bombing there was, naturally, heightened anxiety with respect to the possibility of additional, similar attacks. However, that concern does not translate into instinctively presuming that actions that would otherwise be ignored or deemed harmless justify overreaction. Simply put, in the immediate aftermath of an attack, government engages in excess with respect to similar *modus operandi* (in this case, air travel) by assuming that a similar attack is in the offing. While doing so may serve to calm a jittery public whose confidence in government has lessened, given the obvious failure that enabled the attack to occur, this is basically looking outside to see a sunny day and assuming that, since it looks nice, tomorrow will likely be sunny as well. In so overtly focusing on air travel, and specifically targeting the prevention of attack modes that have already been tried, government incurs significant risk by both overcommitting resources to a particular target and minimizing protection with respect to other, high-probability targets or methods.

In the immediate aftermath of Flight 253, administration officials focused exclusively on air travel and the need to expend significant resources to protect those traveling, at the cost of ignoring other threats. President Obama was quoted as stating that the bombing reflected a "screwup" (a polite understatement for the reality was a colossal failure); he is right, of course. But, the appropriate response is *not* to immediately and instinctively engage in outrageous expenditure (body scanners); rather, the more appropriate response is to literally take a deep breath, consider alternatives, and respond in a measured manner. For instance, merely talking to passengers in the preboarding process would have prevented the attack. Alas, experience suggests the following: decision makers in response to an attack (regardless of whether it is successful) make three primary assumptions:

1. The next attack will be similar (the adage that "generals fight yesterday's war" is appropriate).
2. The public exclusively—perhaps understandably—focuses on a weakness identified by terrorists to which a response is demanded.
3. Other potentially vulnerable targets take a back seat.

This suggests an important—and possibly inherent—limitation in the cost–benefit analysis: the need to mollify the public at the expense of protecting the public. That is, rather than adopting measures less dramatic than overkill (scanners), *apparent* minimization, rather than excessive (and expensive) maximization, would significantly contribute to rational-based cost–benefit analysis. There is, however, a significant tension in this discussion: public expectations, following an attack, demand excessive show and response.

Herein lies a crucial dilemma for decision makers: effective homeland security is negatively impacted when government responds to what is perceived and defined as public pressure. In many ways, the cliché "less may be more and more may be less" illustrates how excessive response fundamentally skews cost–benefit calculation, thereby negatively impacting quantifiable effectiveness.

Put another way: The fact there may never be another attack on a commercial jet liner does not mean that measures imposed after Flight 253 were successful; potentially, quite the opposite. Costs incurred are, after all, at the expense of responding to—and minimizing—*other* threats. The danger of overreaction, however, is significant precisely because it directly impacts how decision makers potentially protect against—and respond to—additional, viable threats. Which brings us back to threats.

31

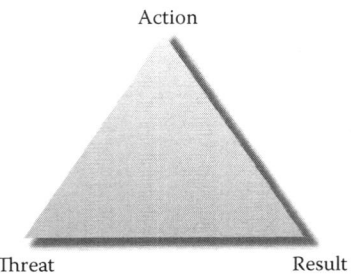

FIGURE 2.5 Triangle of action, result, and threat.

To wit, the following are representative of overreaction with attendant costs, minimal benefits, and dubious contribution to homeland security, much less passenger safety. As Professor John Mueller noted:

> Sometimes security measures can even cost lives. Increased delays and added costs at airports due to new security procedures provide incentive for many short-haul passengers to drive to their destination rather than flying. Since driving is far riskier than air travel, the extra automobile traffic generated by increased airport security screening measures has been estimated to result in 400 or more extra road fatalities per year.*

In a telling email a TSA official wrote to me:

> I agree that our threat mitigation should be improved. Take the recent security breach [referring to Newark Airport] where they had to shut the airport down which delayed flights across the globe.†

The official is, of course, correct; the common denominator of the above examples is a response devoid of sophisticated threat analysis and determination. The breach (undoubtedly significant) resulted in a shutdown of a major international airport for six hours affecting 12,000 individuals. The question is whether there was a direct connection between the action (breach), the result (lockdown), and the threat (direct or perceived) posed (Figure 2.5).

The essence of effective and efficient cost–benefit analysis requires a sophisticated understanding of the triangle's three legs separately and

---

* http://psweb.sbs.ohio-state.edu/faculty/jmueller/ISA2008.pdf (citing Jerry Ellig; last accessed January 9, 2010).
† Private email; in author's records.

collectively. The question, in the cost–benefit context, is whether TSA officials responded according to standing orders predicated on previously determined simulated scenarios and viable, time-relevant threat assessments, or whether the response reflected panic and disorder.

If the latter, then 12,000 travelers were victims both of the individual who breached the security line (who appropriately faces criminal prosecution) and of federal security officials who failed the public by not properly planning and foreseeing this event. If the former, then the question must be asked whether relevant standing orders were sufficiently nuanced and fine-tuned. After all, effective cost–benefit requires minimizing the former while maximizing the latter.

This is easier said than done, for the intervening variable is developing and implementing a proper understanding of possible threats. In the context of managing or minimizing terrorism—compared to false and ludicrous claims of defeating terrorism—the government has limited resources with which to achieve limited goals. For that reason, the state cannot "bring the cavalry" to every threat, nor should it. The question of when to bring—and how much—is the fundamental operational dilemma to be considered in the cost–benefit discussion. The tension from a decision maker's perspective is to correctly gauge both considerations when public and media are clamoring that "by all means necessary" be the appropriate philosophical and operational paradigm.

While attractive from a public relations perspective, reality suggests otherwise; the state has neither unlimited resources nor unlimited legal and moral boundaries in responding to either threat (perceived or actual) or attack. How to allocate limited resources and assets is extraordinarily difficult. It is also essential to implementation of effective homeland security.

By way of example: In response to a perceived Iraqi threat to attack Israel with missiles, my colleagues and I serving in the HFC were subjected to an endless series of meetings that would drag on incessantly. The number of participants ranged from intimate numbers to hundreds of officials; the cost in time and resources was significant, whereas the benefit significantly decreased in direct proportion to the number of meetings and number of participants. Simply stated: The more meetings scheduled, the less productivity both individually and collectively. The same is true in the United States. A recent *Washington Post* survey found that some 1,271 government organizations and 1,931 private companies work on programs related to counterterrorism, homeland security, and intelligence.* Additionally, an estimated 854,000 people hold top-secret security clearances.†

There is an important lesson here: one of the realities of homeland security is that threats, risks, and dangers are largely murky and unarticulated. They are certainly moving targets, suggesting a lack of clarity and conciseness in threat definition and identification. The result of that reality is, invariably and inevitably, overreaction and the risk of misguided responses from a tactical and strategic perspective alike. Ineffective and ineffectual cost–benefit analysis—whether panic-based lockdowns or resource mismanagement—reflects a critical flaw regarding homeland security.

The failure to clearly articulate both what presents a viable threat and what resources to dedicate in order to mitigate that threat (loosely defined) directly leads to a failure to carefully analyze threats and their possible consequences from a cost–benefit perspective. Simply put, the response to a homeland security crisis (whether real or perceived) is generally excessive. The harm, in the context of air travel, is significant from the perspective of time and money alike.

On the other hand, some will argue that *any* threat justifies overreaction; otherwise, so goes the argument, innocent civilians will unduly be placed in harm's way—a risk government cannot take. Were resources unlimited (financial, time, personnel), then a full-scale response to every threat (perceived or real) would be possible, albeit unjustified and unnecessary.

Whether that would objectively increase homeland security is, however, an open question. I would suggest such an approach—clearly unfeasible—is also ineffective. That is, such an approach would not increase either individual or collective security and would directly affect the society we are. A policy of placing a police officer at every street corner (whether figuratively or literally) risk the following significant "collateral harms:"

---

* http://projects.washingtonpost.com/top-secret-america/articles/a-hidden-world-growing-beyond-control/ (last accessed October 24, 2010).
† Id.

1. Limiting individual freedoms (increased possibility of stop and frisk searches and other possible unwarranted invasions of privacy by overcautious officers)[*]
2. Viewing all threats with equal severity, therefore failing to discern trees from forests[†]
3. Inability to satisfactorily respond to non-homeland security threats (basic law enforcement responsibilities)[‡]
4. Misallocation of resources required for other, non-homeland security-related challenges facing society (i.e., education and health)

---

[*] http://www.freep.com/article/20100117/NEWS05/1170511/1319/Deadly-FBI-raid-of-Detroit-mosque-prompts-concern-over-informants; http://www.msnbc.msn.com/id/34846903/ns/travel-tips (last accessed October 24, 2010).

[†] http://www.nytimes.com/2010/01/16/nyregion/17jfk.html?hp (last accessed October 24, 2010).

[‡] http://www.nytimes.com/2010/01/17/sports/olympics/17border.html (last accessed October 24, 2010).

However, precisely because the reality is that society cannot afford (by any measuring stick) to place the police officer at every corner, the essence of the discussion is resource allocation, prioritization, and threat assessment. Otherwise, cost–benefit analysis is bereft of any substance and the response by decision makers is not predicated on sophisticated assessment of society's points of vulnerability. To truly engage in the cost–benefit discussion, decision makers must adopt an approach defined as "bean to cup." That term of art means the following: points of vulnerability (regardless of the enterprise/entity) are carefully categorized and delineated, thereby facilitating determination regarding threats and required protection levels. All societies have particular points of vulnerability; the challenge—in the "bean-to-cup" paradigm—is realistically assessing which points of vulnerability demand particular protection (and degree) and which potential targets are less vulnerable, and therefore less demanding of protection.

Otherwise, the cost–benefit analysis is devoid of empirical analysis that significantly facilitates more effective asset protection. If the primary objective of homeland security is to maximize asset protection (however society defines assets), predicated on the most prudent and efficient allocation of resources, then objective cost–benefit analysis is essential.

Resources are limited; given the current economic crises, there is a need to engage in critical discussion addressing fundamental rearticulation of resource allocation. That is, in many ways, the essence of cost–benefit analysis: while society would like to believe and decision makers would like to convey that *all* assets are equally protectable, that is, at best, illusionary.

In the aftermath of 9/11 there was, by example, an understanding that chemical plants must be protected. Sounds good, and even makes sense. But the questions attendant to such an overarching policy are no less important than the policy itself. That is, we must—in the context of cost–benefit analysis—ask the following questions:

1. How are standards determined? (That is, are the standards—and their inevitable effect—arbitrary and capricious?)
2. How should costs be absorbed? (industry, government, consumer)
3. What are threat levels relevant to the industry?
4. What risks can be assumed? (with understanding that not all threats can be mitigated)
5. What impact do standards have on industry?
6. How have threats been identified and categorized? (predicated on the assumption that threat categorization is inherent to developing standards)
7. Are less burdensome alternatives available? (i.e., cost, manpower)
8. What is the desired effect/outcome of standards? (That is, are they merely palliative or are concrete results expected?)
9. What are matrices for assessing the effectiveness of standards imposed on industry?
10. What are side effects of imposing standards on industry? (That is, how could resources otherwise have been spent/allocated, and what is the cost of not doing so?)

The "bean-to-cup" approach, if properly implemented, facilitates both threat assessment and measured response clearly lacking in the debate. Basing cost–benefit analysis on "bean to cup" significantly enhances rational-based homeland security that would—in the short term and long term alike—more effectively protect innocent civilians. While not all assets are protected (nor should they be), and those

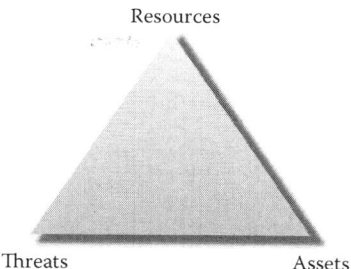

FIGURE 2.6   Triangle of resources, threats, and assets.

protected (predicated on thorough analysis) would not be protected to an equal degree, the "bean-to-cup" approach enables a strategic confluence between resources, threats, and assets (Figure 2.6). This confluence would enable the most effective articulation of cost–benefit analysis, directly leading to effective homeland security.

By not adopting such an approach, decision makers will continue to respond in accordance with the traditional panic response. That well-worn model is the epitome of ineffectiveness, inefficiency, and profound misallocation of valuable, limited resources. It also, as suggested in Chapter 1, unnecessarily endangers people. After all, the inexcusable manner in which the lockdown of Newark International Airport was conducted— with its attendant swarming crowds rendered extraordinarily vulnerable to either a sophisticated terrorist attack or panicked stomping—provides the most compelling (and graphic) justification for the "bean-to-cup" approach.

What is most troubling is the failure to systemically engage in a sophisticated "lessons learned" approach to cost–benefit analysis of homeland security. Otherwise, the responses illustrated in this chapter would not represent a recurring theme best described as systematic over-response that results in unnecessary expenses coupled with minimal benefits. While some might hearken back to Benjamin Franklin's wise words (as suggested at the chapter's beginning), I suggest a more practical approach firmly rooted in rational-based cost–benefit analysis. While doing so requires decision makers to speak truth to the public (regardless of its attendant political complications) regarding the inherent limitations of systemic, mature homeland security, there is, frankly, no alternative. Otherwise, the public will be under the impression (illusion) that, figuratively, the police officer is at every corner and that resources are unlimited.

The reality is far different for resources are far from unlimited, and not only is the officer not at every corner, but more often than not, neither the police officer nor command fully knows where he or she should be. That is a direct reflection of failing to correctly assess threats, resource allocation, and prioritization. Otherwise, how do decision makers explain the pictures from Newark Airport and Northwest Flight 253?

Cost–benefit analysis predicated on the "bean-to-cup" points of vulnerability approach would directly address these core weaknesses.

3

Prioritizing Risks, Threats, and Dangers

"You can't do it all" could be the title of this chapter. The reality of homeland security—regardless of DHS's mandate—is that it is impossible to realistically demand or expect that all 18 critical infrastructures be uniformly, constantly, and effectively protected. Limited resources—financial and manpower—dictate that inherent limitation. Therefore, the magical—and operative—phrase is "prioritization of resources." How to accomplish this essential goal will be our focus in this chapter. The prioritization discussion must be understood in the larger context of how to most effectively protect the homeland; as Figure 3.1 and Figure 3.2 depict, an integrated, multidisciplinary approach is essential.[*]

Undertaking this discussion requires that decision makers articulate to both themselves and the public that *limitations* is the critical byword to a sophisticated, sober discussion regarding homeland security. Twenty years at the operational counterterrorism table convinces me of the absolute correctness of this recommendation. To best facilitate operationalizing prioritization, this chapter includes a number of checklists intended to facilitate discussion and debate.

The checklists will be our guide in addressing and analyzing prioritization. In many ways, understanding the reality of homeland security

---

[*] Amos Guiora, International Cooperation in Homeland Security, available at http://papers.ssrn.com/sol3/papers.cfm?abstract_id=1160067 (last accessed October 24, 2010).

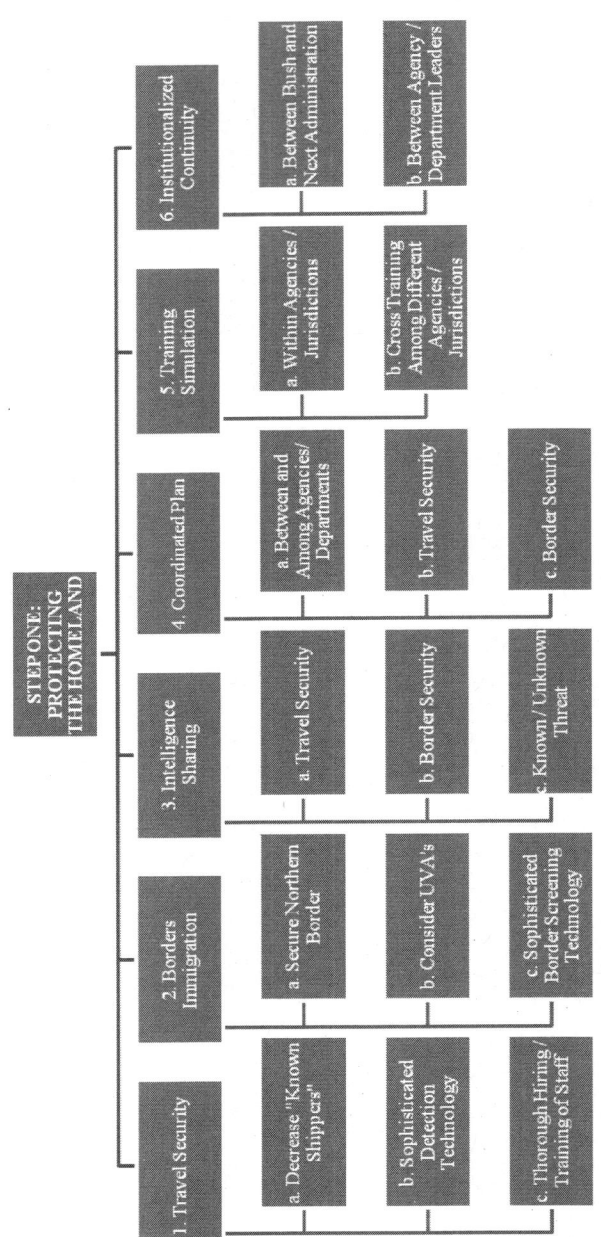

FIGURE 3.1 Step 1: Protecting the homeland.

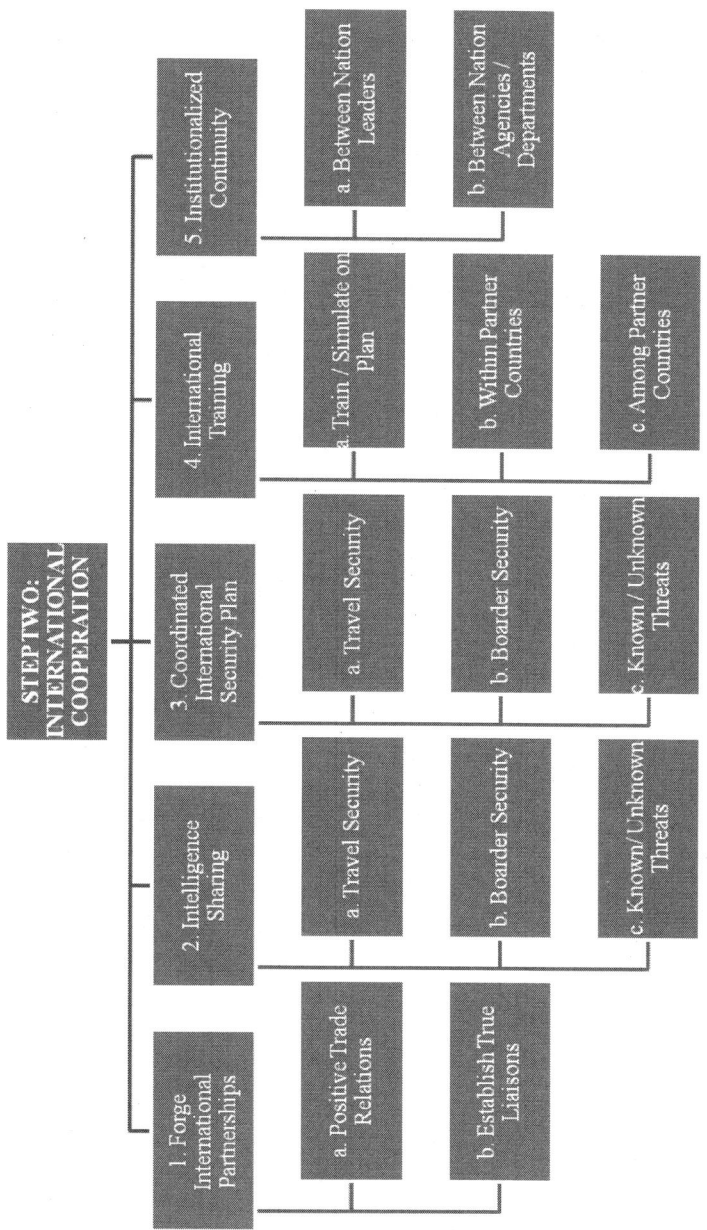

FIGURE 3.2 The threat imminent continuum.

requires understanding the requirement to prioritize. Otherwise, the limits of homeland security will not be addressed. To dismiss the limits of homeland security is both intellectually dishonest and practically incorrect for it also creates a dangerous illusion among the public. Prioritization is determining how to allocate limited resources when attempting to achieve a specific goal. In the homeland security paradigm it means decision makers must determine *how* to allocate limited resources while saddled with an *unrealistic* mission statement. A candid discussion regarding prioritization suggests how to address—if not resolve—this problematic conundrum. To that end, the checklist below articulates basic prioritization principles.

Prioritization is dependent on the following factors:

1. Articulation/determination of threat (perceived or actual)
2. Articulation of available resources (in determining what, how, and when to protect)
3. Articulation of *realistic* homeland security goals
4. Determination (by decision makers) of what assets are expandable/marginal
5. Risk assessment regarding danger of not protecting particular asset
6. Political considerations (fallout) attendant to not protecting particular asset
7. Perception of prioritization by other side
8. Media spin and public perception of prioritization
9. Articulation of economic considerations (were unprotected target to be attacked)
10. Articulation/determination of training/expertise required (including costs) to effectively protect assets
11. International cooperation, which facilitates effective prioritization, particularly with respect to intelligence and information sharing, resource sharing (as an example, joint training exercises), and exchange of lessons learned*

---

* For a fuller discussion, see Amos N. Guiora, International Cooperation in Homeland Security, available at http://papers.ssrn.com/sol3/papers.cfm?abstract_id=1160067.

THREATS

Threat analysis facilitates decision makers' ability to determine what asset is most directly at risk. Protecting all assets uniformly is beyond the operational, financial, and practical capability of any homeland security agency; by engaging in a sophisticated and realistic threat analysis, decision makers can most prudently determine how to effectively allocate resources. However, it is a reality that the combination of prioritization, cost–benefit analysis, and risk assessment does not guarantee absolute homeland security. What an integrated approach seeks to accomplish is to minimize harm to society. The danger, particularly in response to either a direct or perceived threat, is overreaction, which potentially violates otherwise constitutionally guaranteed protections while neither effectively or efficiently protecting the public.

Threats can best be visualized through the graphic shown in Figure 3.3.[*]

Threat assessment, then, must be rational-based, reflecting analysis of:

- Intelligence information (determination of whether intelligence information is actionable depends on assessing whether it is reliable, viable, valid, and corroborated).
- Attack trends (whether successful or thwarted).
- Points of vulnerability (see Chapter 2 for the discussion regarding bean to cup).
- Effectiveness of existing homeland security measures (relative to financial and manpower resources allocated to specific measures).
- Viability and credibility of previous threats emanating from similar sources; this is measured by assessing whether an actual attack occurred, what measures—if any—were taken in response to the threat, the cost of those measures, and how, if at all, the threat was presented to the public.
- How the risk of not responding to a previous threat was assessed in the context of potential harm to assets deliberately not protected and to assets targeted by the threat (whether perceived or direct).
- Assessment of previous responses (including nonresponses) to similar threats; the assessment must include multiple variables, including loss of life, property loss, effect on society, and perceived gain for attackers.

---

[*] http://papers.ssrn.com/sol3/papers.cfm?abstract_id=1090328 (need to fully cite the article).

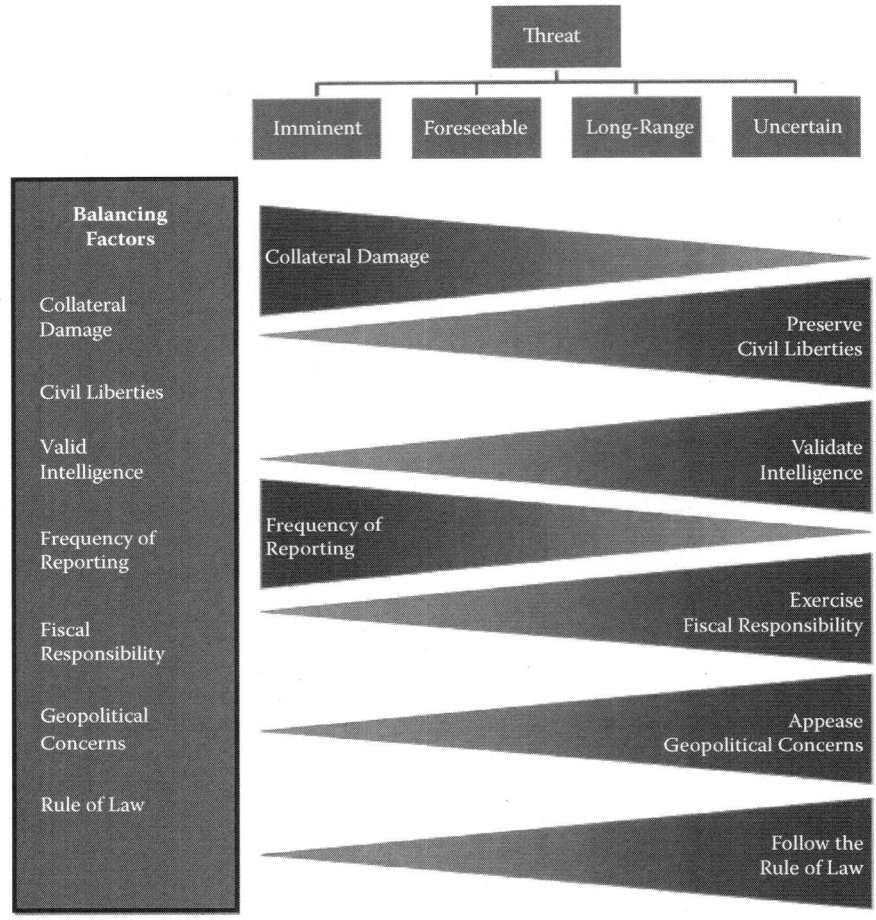

FIGURE 3.3 Step Three Threat Assessment. (Amos N. Guiora, Fundamentals of Counterterrorism, New York: Aspen Publishers, 2008, 153.)

- Current state of homeland security (i.e., recent attacks/threats; major events scheduled that invite attacks, including Super Bowl, Olympics, presidential conventions; geopolitical developments that impact domestic decision making; resources allocated by Congress).
- Political, social, and economic realities of communities that threats would impact should they come to fruition.

- First responder capability in relevant communities (this requires analysis of resources allocated; demonstrated technical and operational abilities; previous experiences; skills training, including integration of firefighters, paramedics, law enforcement, and volunteer organizations; and defined scope of areas of responsibilities).
- Private sector–public sector cooperation level*; the resilience of the home front is largely dependent on cooperation between the two sectors. Akin to the requirement that federal law enforcement share intelligence information with local law enforcement (see point 13 below), the former must do the same with the business community (regarding employees,† threats, risks, and capabilities).
- Operability of business continuity in face of threat effectuation. (For example, a major international corporation based in Cleveland, Ohio, conducted a large simulation exercise intended to determine its preparedness levels with respect to terrorism and natural disasters. In the debriefing, senior executives were chastised for having ignored basic items in a self-imposed quarantine scenario, including the need to provide food, water, and backup generators.)
- Local-state-federal law enforcement cooperation level. While the 9/11 Commission admonished various federal agencies for a failure to communicate (particularly the FBI and CIA), the requirement for federal and local cooperation is not critical, particularly with respect to intelligence information/assessment. However, as the following vignette illustrates, implementation is extremely complicated: at a meeting with FBI agents I suggested that successful homeland security requires that the bureau view local law enforcement as equal partners; their response can only be described as apoplectic.

What this proposed checklist highlights are the issues that must be examined in an integrated, multidisciplinary fashion to most realistically reflect the enormous complexity of homeland security. A realistic assessment of resource allocation, based on the points above, will directly

---

* See my congressional testimony on this issue, available at http://www.fas.org/irp/congress/2008_hr/051508guiora.pdf (last accessed October 24, 2010).

† Amos N. Guiora, Counterterrorism and Employment: An Israeli Perspective, available at http://papers.ssrn.com/sol3/papers.cfm?abstract_id=785368 (last accessed October 24, 2010).

contribute to effective homeland security policy. Reliance on a silo (nonintegrative) threat model rather than an integrated model capable of analyzing different—and competing—variables significantly hinders homeland security. Competing variables are critical to understanding homeland security; in many ways, homeland security reflects a moving target with many elements unseen. By analogy, I have often referred to terrorism as the quintessential reality of unseen enemy with dark shadows in back alleys.

Effective threat analysis—particularly when resources are limited—is extremely complicated because intervening variables cloud rational decision making. Without doubt, conveying that inherent ambiguity and murkiness to the public presents extraordinary challenges to decision makers. On the other hand, given economic and operational realities, decision makers have no choice but to implement a policy predicated on prioritization and articulate that directly and candidly to the public. While admittedly difficult and, in many ways, antithetical to politicians, homeland security demands straight talk, devoid of false expectations predicated on illusions and falsities.

Prioritization is the essence of that philosophy because it clearly reflects the limits of governmental capability and defines the limit of government's reach. However, decision makers are extremely sensitive to how traditional and nontraditional media alike analyze prioritization and risk assessment. The reality of the 24/7 news cycle (particularly the blogosphere) dramatically impacts flexibility craved by decision makers. The transformation of how Americans receive their news—from Walter Cronkite to cable to the Internet—directly impacts homeland security strategy. Fear of public reaction suggests that the safe course is to uniformly protect all assets, regardless of their verifiable and actual strategic importance. In reality this is a falsity for neither can all assets be protected nor can those that are protected be equally protected.

Not only have decision makers largely lost the ability to control how (and when) news is disseminated, but also the sheer number of self-anointed commentators and pundits directly impacts how critical issues are presented and articulated to the public. Frankly, it is low-hanging fruit to criticize decision makers for failing to effectively protect (i.e., prevent an attack) an identifiable asset. It is much more difficult for decision makers to explain (even when correct) why the attacked asset was unprotected, and that the decision to not protect the asset was predicated on a rational analysis.

This reality unquestionably clouds effective homeland security strategy. In a less venom-filled political culture, prioritization of threats (direct and perceived) would be the centerpiece of homeland security policy. However, political realities drive the discussion; therefore, the critical question before decision makers is how to utilize the checklist above while articulating the following fundamental essence: *some assets will be unprotected*. What complicates the discussion is the operational impact of prioritization-based homeland security: certain risks will deliberately not be averted or minimized precisely because resources are otherwise allocated and therefore unavailable. Prioritization, doubtlessly, requires decision makers make unpopular choices that can lead to the loss of innocent lives. The essence of the paradigm, then, is minimizing unknown risks while fully understanding and appreciating that *uncertainty* is the critical variable when undertaking resource-driven risk-based homeland security.

The inevitable visceral reaction to this reality presents significant challenges to decision makers. The checklist unequivocally manifests the inherent limitation of homeland security. However, by creating a systematic mechanism facilitating a rational-based approach to prioritization, the impact of the inability to protect all assets is minimized. What the checklist enables is risk minimization based on careful analysis of threats and resources alike. To do this requires understanding the nature of the threat both tactically and strategically. That is, the threat (received from a source) must be analyzed as to both its viability and a determination of which available resources to allocate in response to the threat. Those two considerations are tactical; the strategic question is: What are the long-term ramifications were the threat to be consummated and decision makers had (predicated on a rational-based analysis) determined not to allocate what, in retrospect, seems obvious?

The long-term implications/ramifications must be viewed from the perspective of multiple audiences; much like prioritization cannot be viewed through a silo (but rather through an integrated analysis), the impact of a successful (from the perspective of the actor) attack must be analyzed from the following perspectives:

1. **The general public:** The public will not be understanding of rational-based resource allocation that left assets unprotected even if the assets' significance is a legitimate matter of public opinion and debate; the fact that a successful attack occurred is sufficient in and of itself, regardless of the impacted assets' true significance both tactically and strategically.

2. **Decision makers:** They will face unrelenting media scrutiny if a threat not to be responded to resulted in loss of innocent life and property damage; the challenges are threefold:
   a. Create, implement, and articulate rational-based criteria for threat assessment.
   b. Explain that the essence of prioritization suggests just that: not all threats can be mitigated and not all attacks prevented.
   c. Articulate the benefits of risk assessment in developing a prioritization policy.
3. **Terrorists:** Terrorist leaders are extraordinarily sensitive to the rationale guiding decision makers. (On a personal note, I was always impressed with the sophisticated understanding of both Israeli policy and decision makers that Palestinians articulated; more often than not, I found myself thinking that their insight of us was deeper than our insight of our decisions and decision-making process. To that end, it is important to recall the adage "the occupied speaks the language of the occupied; the occupier does not speak the language of those occupied.") The decision to protect or not protect a particular asset will be carefully analyzed by nonstate actors (always, it is important to add, with inherent suspicion and skepticism regarding the decision makers' motives). Furthermore, a successful attack will have enormous importance for the nonstate actor, as it reaffirms both (from the public's perspective) their operational capability and the state's commensurate weakness and vulnerability. To that end, an unsuccessful attack (i.e., Northwest Flight 253) is still of enormous value from the perspective of the nonstate actor, given that state incompetence and vulnerability were clearly exposed and the resulting political and media noise suggests a fundamental lack of confidence in state decision makers.
4. **First responders:** They are clearly impacted both from a failure to properly assess the viability of an actual threat and from a prioritization policy that enhances their vulnerability. While first responders' dissatisfaction with prioritization/resource allocation decisions is understandable, policy makers must distinguish between a tactical response (the essence of first response) and strategic decision making (the fundamental responsibility of state decision makers). First responders are, in many ways, the foot soldiers Tennyson so vividly described when he wrote: "Ours is not

to wonder why; ours is but to do and die.'" Policemen, paramedics, and firefighters risk their lives to save ours; they are, indeed, the heroic foot soldiers of homeland security, for they respond, no questions asked. While resources required to fully respond to natural disasters and terrorism alike will be inherently wanting, the obligation of first responders is to protect the community with available resources. This is obviously not the ideal from the perspective of public and first responders alike, but it is the purest manifestation of prioritization. In many ways, the prioritization discussion can be most succinctly examined through the lens of first responders: Was there sufficient equipment to protect and respond to terrorism and natural disaster?

5. **Media:** Predicated on my professional experience (IDF and academia), the media will articulate a successful terrorist attack as state failure and will not accept an argument that the state had engaged in rational-based decision making regarding prioritization and resource allocation. As discussed above, this presents a major challenge to decision makers; while unavoidable (given the realities of the communications age), it presents complex challenges that few decision makers address competently and comfortably.

While the inability to protect all perceived assets is an operational reality (though politically an extraordinarily difficult sell), the more appropriate question is: How do we most effectively and strategically allocate *actually available* resources? That is, the essential mission (protecting the public) must be examined differently than presently articulated by decision makers who make fundamental misrepresentations to the public regarding DHS's operational capability. In other words, protection must be based on prioritization and resource allocation relying on concrete risk and threat assessment models rather than presumed abilities. Bottom line: How do decision makers truly decide *how to* allocate resources? The basic requirement, as illustrated in the following checklist, is engaging in an effectiveness analysis; after all, prioritization must reflect careful analysis of how to effectively (with limited resources) create a successful homeland security policy.

To that end, to determine whether a measure is effective, the following questions must be asked:

---

* Alfred, Lord Tennyson, The Charge of the Light Brigade (1854).

1. Is the measure preventative, preemptive, or retaliatory?
2. What is acceptable collateral damage? (This question is relevant to operational homeland security.)
3. What are the financial costs?
4. What are the costs to civil liberties?
5. What are the geopolitical costs?
6. How valid is the intelligence?
7. To what extent does the measure follow the rule of law?
8. What alternatives exist?
9. To what extent does the measure overlap with existing measures?

DHS is analyzed by the Office of Management and Budget's (OMB) Program Assessment Rating Tool (PART). According to PART, 42% of surveyed DHS programs rated either effective (14%) or moderately effective (28%); however, 25% of surveyed programs rated only adequate. The remainder, nearly 33%, yielded a "results not demonstrated" classification. Using a similar PART rating system, OMB determined that 13% of DHS spending (representing $5.4 billion) was effective, 53% ($22.2 billion) was moderately effective, and 19% ($8 billion) was merely adequate. Fifteen percent of DHS spending ($6.2 billion) was counted as "results not demonstrated."

While it is indisputable that progress has been made within the Department of Homeland Security, public criticism often contradicts the optimistic finding of the PART surveys, indicating that much remains to be done before all DHS programs can be considered effective. Lack of organization, resources, and leadership contributed to DHS ranking last in job satisfaction among 36 surveyed government agencies last year. Homeland security experts claim the department's emphasis on law enforcement and security rules, as opposed to greater focus on intelligence and nuclear nonproliferation, is ultimately undermining the nation's security from terrorist threats. Many experts cite the failure of Congress to consolidate its oversight of Homeland Security (ignoring the 9/11 Commission's recommendation to do so) as a major impediment to effective action. Indeed, Secretary Chertoff often cited the difficulty of protecting the nation while serving so many masters with inconsistent positions.

I suggest the following model: referring to the Israel-Palestinian conflict, former IDF chief of staff Moshe (Bugi) Ayalon described "the battle of the narrative" as extraordinarily critical. In the context Ayalon was discussing, he was commenting that while operational counterterrorism measures are of the utmost importance, public perception is no less important. In the prioritization–threat assessment discussion public perception is critical for the public demands/expectations that all assets be protected, but that is frankly impossible. The question, then, is how do decision makers resolve these critical tensions and more effectively balance limited resources while both protecting (strategically) the public and directly addressing public perception?

By developing a workable model based on the checklist above, prioritization of limited resources based on threat analysis will be significantly enhanced. To that end, decision makers must articulate the following: while it is impossible to protect all assets, threats deemed viable justify protection, and measures implemented will afford maximum protection. Rearticulated, the decision to allocate resources to protect assets must reflect sophisticated, strategic analysis.

PRIORITIZING THREATS

To meet this critical goal, decision makers must ensure that policies and practices facilitate appropriate and effective responses to threats (direct and perceived). By example: The egregious failure to prevent Abdulmutallab from boarding Northwest Flight 253 is only matched by frightening incompetence in the immediate aftermath of the unsuccessful attack.[*]

Prioritization, then, is essential for resource allocation with respect to threats both prior to an event (determining which assets to protect) and during an actual event. Akin to triage in a medical emergency context, sophisticated prioritization analysis facilitates actual resource allocation. That is, by engaging in a realistic discussion of means available to decision makers, it is possible to develop a policy that mitigates both the damage from the initial event and its possible, additional, fallout. Prioritization in the immediate aftermath of an attack requires practiced restraint, for the instinctual response is to "send in the cavalry." However, because an additional attack is always a possibility, husbanding of resources is essential to effective homeland security.

By example: If initial reports indicate that terrorists have attacked a particular school, decision makers must prioritize their resources to ensure that a sufficient number of first responders are available for a

---

[*] http://www.freep.com/article/20100129/NEWS05/1290375/1321/Response-botched-after-Northwest-bomb-attempt (last accessed January 30, 2010).

possible second attack. Practically speaking, if all available first responders (i.e., fire, paramedics, and police) travel in the same direction, clogged roadways are an inevitable result. Should there be a second attack and all units responded to the first attack—in accordance with misguided emergency response plans—then redirecting (i.e., "U-turning") will be a logistical nightmare. It will also, tragically, cost lives.

I have witnessed this both as a participant and an observer; the following two vignettes are indicative.

When serving as the legal advisor to the Gaza Strip I received a request on my pager to arrive ASAP at a particular location in the Gaza Strip where a suicide bombing had occurred. While only the bomber had been killed (the device malfunctioned), the *modus operandi* was most unusual (the terrorist was on a donkey-led wagon), which attracted (because of its uniqueness) numerous curious onlookers; the result was that many IDF officers had been ordered to arrive (the road was immediately closed to Palestinian residents of the area).

We (the responding IDF officers) were clearly vulnerable to an additional attack; finally, the commanding officer ordered all those who did not have clearly defined operational responsibility to immediately leave the area. While his decision was correct, it was, frankly, late in coming. In other words, if the failed attack had been part of a more sophisticated terrorist endeavor, then the commander's decision could have been best described as a day late and a dollar short.

This was not an irrelevant consideration, for in 1995, 20 IDF soldiers were killed in a double bombing: in the immediate aftermath of the first bombing, soldiers rushing to aid the wounded were killed in a second, preplanned bombing. Precisely because their response was instinctual—and therefore predictable—lessons learned would emphasize the need to dilute (perhaps minimize) the number of responders to the initial attack. This is the operationalizing of resource prioritization; it is also essential in developing a coordinated approach to threats and attacks alike. Doing so requires decision makers carefully scrutinize the relationship between resources and the threat or event that the former are intended to address, if not mitigate. That is, resource allocation must be specifically dedicated to a particular threat; otherwise, spending and response will equally reflect what can best be described as "catch as catch can." The danger posed by secondary threats particularly highlights this necessity for failure to adequately prepare for such an attack inevitably leads to overresourcing for the initial attack, and then significantly hindering the ability to sufficiently respond to the second attack.

While clogged streets—impassable to citizens and law enforcement alike—are the all but predictable result of attacks on the civilian population, effective planning and prioritization significantly minimize gridlock (practically and intellectually). To achieve this critical goal, decision makers, in addition to the previous checklists, must address the following:

1. Were a school (i.e., high-visibility/high-impact target) attacked, what are the most obvious traffic patterns (from the perspectives of parents and law enforcement), and what then would be possible (based on threat analysis and intelligence information assessments) secondary targets that would directly limit response?
2. How many resources are needed to respond to an initial attack (predicated on intelligence analysis of a potential attacker's capabilities)?
3. What are state's realistic (requires complex self-assessment) operational capabilities?
4. What are the relevant population group's self-control/self-restraint capabilities (requires complex psychological analysis of particular communities)?
5. When assessing points of vulnerability (i.e., shopping malls, major corporations, industrial zones, "main street"), decision makers must both analyze possible ramifications (i.e., costs) of an initial attack and identify potential secondary targets (i.e., additional points of vulnerability) and determine cost of inadequate protection.
6. In determining the state's possible operational response to the initial attack, it is important to calculate possible costs attendant to insufficient capability with respect to secondary attack.
7. What are foreseeable financial losses in the event of multiple attacks (initial and secondary)?
8. In the context of secondary threats/attacks, decision makers must prioritize logistic/transportation-related assets (i.e., roads, bridges, airspace, harbors, emergency vehicles, airports).

9. The possibility of secondary attacks requires predesignation of medical assets in order to avoid overburdening individual hospitals/medical centers.
10. The predesignation of zones of responsibility for law enforcement based on secondary threat assessment must be determined in a coordinated manner.

In the event of a secondary attack, failure to have predesignated resources will endanger the secondary target more than the actual initial target. Careful prioritization and sophisticated resource allocation significantly minimize otherwise avoidable loss of life and property. However, common sense and political realities suggest this process is enormously complicated. When former speaker of the U.S. House of Representatives Tip O'Neill (D-Mass) famously quipped that "all politics is local," he could have been referring to the dilemmas suggested above.

By way of example: Shortly after I was first appointed to be the HFC legal advisor, Saddam Hussein became increasingly belligerent regarding Israel. Actually, Hussein constantly and repeatedly threatened to attack Israel, suggesting availability of and willingness to employ unconventional warfare. The result was significant concern among the Israeli population and realization among decision makers that significant resources had to be allocated to provide the home front with sufficient protection. What was missing, perhaps resulting from legitimate concern, was careful discussion regarding resource allocation and limits of responses. While the intended result was to assure the public that no efforts had been spared (a highly problematic—albeit conceivably necessary—message), fundamental questions regarding prioritization were left unattended.

Simply put, the response was intended to demonstrate that matters were under control, but reality suggested otherwise. The primary reason for this is that resource allocation and asset prioritization were not discussed. A sophisticated public understands (instinctually and practically) that limits are an inherent aspect of homeland security; a failure to engage in this critical discussion fundamentally disserves the public. While there was, among those responsible for home front security, enormous willingness to serve the public, critical ingredients were either unaddressed or addressed on the fly. The following were issues warranting further discussion; engaging in prioritization and risk assessment discussion would have enormously facilitated their resolution.

1. What organization bore fundamental, core command responsibility for all aspects of home front security?
2. What potential targets were deemed expendable and therefore warranted less than maximum protection?
3. What resources were to be expended on direct protection of the civilian population (i.e., when and to whom were gas masks to be allocated)?
4. How were different/distinct first responder organizations to coordinate their efforts?
5. What is the responsibility of local government?
6. What is the obligation of schools to educate and protect children during a homeland crisis?
7. When should national leaders address the public and what should be their fundamental message?
8. What decisions should be made with respect to alternative uses of existing infrastructure (i.e., what buildings can serve as shelters for those whose homes and property have been destroyed)?
9. What is the proper role of volunteer relief organizations?
10. What can be learned from the experiences of other countries?

Some of the questions above must be addressed through a local lens; others can be examined through a truly international angle. What is important—whether the perspective is local or international—is that cost–benefit analysis, resource allocation, and prioritization are principles essential to articulating and implementing effective homeland security. Government resources are limited; conversely, the threats posed (direct or indirect) continue to grow exponentially. As demonstrated on Christmas Day, efforts are not spared to attack innocent civilians, regardless of their ethnicity, gender, age, and national origin. The essence of terrorism, after all, is its inherent randomness with respect to how terrorists define legitimate targets.

However, as discussed in this chapter, the state has both limited resources and limited intelligence (gathering); the philosophical and practical underpinning of prioritization is that there are difficult days ahead, as not all possible attacks can be prevented. To suggest otherwise is to deceive the public while expending enormous financial resources. This is a luxury that, today, is untenable and cannot be prolonged.

The checklists proposed in this chapter are intended to further the prioritization, cost–benefit, and resource allocation discussion. Without this discussion we will continue to—literally and figuratively—throw money away. It is only hoped that we do not throw the baby out with the bathwater.

4

International Cooperation, Intelligence Gathering, and Threat Assessment

Air travel has become loathsome and we now avoid it at all costs. The TSA's policy of reactively implementing ineffective countermeasures in order to maintain the completely false illusion of airline security at the expense of the traveling public is profoundly affecting our ability to travel effectively. Example: The "advanced" screening equipment after the Christmas Day bomber is unnecessarily intrusive, offensive and does nothing whatsoever to mitigate the threat posed by the Christmas Day bomber.

The systematic lapses by Intelligence and TSA leave the traveling public at the complete mercy of dumb luck. I am grateful for another day each time I depart a flight.

—Anonymous

Terrorism against the United States, post-9/11, reaches far beyond U.S. borders. In order to effectively prevent terrorism in the United States, decision makers must be sensitive to threats and attacks occurring beyond U.S. borders. International security efforts must concretely address critical issues, including travel security, border control, immigration, intelligence gathering and analysis, and terror financing. Doing so requires enormous effort; furthermore, it requires clear articulation of when a threat is viable and what proactive measures are both lawful and effective in minimizing the danger posed by the identified threat. Nation-states face similar threats subject to specific conditions and local realities. International cooperation facilitates homeland security and need not be viewed as a threat to national sovereignty or domestic considerations. Furthermore, international cooperation does not require—particularly in the context of intelligence gathering/analysis—nation-states to endanger sources or facilitate (inadvertently) domestic vulnerability.

That said, for international cooperation to be more than mere lip service reflecting political correctness, there is a need for pooled efforts. Obviously, the extent to which efforts are genuinely pooled is an open question; it is one that decision makers will increasingly be forced to address. While an instinctual response rejecting—or at least minimizing—international cooperation is understandable and perhaps defensible, contemporary reality suggests that cooperation is essential to effective domestic homeland security. The backbone of international cooperation—once a strategic decision is made regarding value and efficacy—is intelligence gathering/analysis and threat assessment intended to facilitate practical measures relevant to a wide range of issues, discussed below. That is, a two-step process is essential: the decision to engage in international cooperation and then its actual, subsequent implementation.

Determining what issues to address with respect to international cooperation requires identifying the essence of homeland security. While disagreement is legitimate, the issues below represent an important cross section of homeland security issues. Needless to say, the list is not complete, nor does it seek to be or is it, realistically speaking, possible to create one.

To that end, the chapter explores comparative efforts at international cooperation in homeland security by examining Canada, Japan, and the EU. In doing so, the chapter both offers recommendations and articulates criteria by which the United States can improve vital efforts at international cooperation in homeland security. A theme essential to this chapter is the wise adage "no one nation has all the answers; no one nation has defeated terrorism or has guaranteed absolute homeland security."

International cooperation must be understood in the context of both limits and necessity, the former because homeland security measures are inherently limited, and the latter because international cooperation is essential to facilitate domestic security. Though there is a certain risk to a comparative approach—benefits unequivocally outweigh the costs. Learning from positive and negative examples implemented by countries facing similar and dissimilar threats alike enhances articulation, development, and implementation of measures intended to protect the public.

To that end, international cooperation forges professional relationships and partnerships among similar government agencies and parallel national institutions; while each nation-state bears ultimate responsibility for protecting both its civilian population and national assets, developing cooperative measures between nation-states facing similar threats is mutually beneficial. By implementing the steps below, the United States can develop, promote, implement, and maintain effective international cooperation. International cooperation to be effective must not be viewed as a buzzword reflecting mere political correctness; rather, decision makers must commit to a philosophy embracing its intrinsic value. The list below highlights measures necessary to establish international cooperation.

Measures necessary to establish international cooperation include:

1. Creating and institutionalizing sophisticated security measures for domestic and international air travel
2. Creating and institutionalizing secure international borders in response to multiple, developing threats
3. Creating and institutionalizing sophisticated intelligence gathering and sharing mechanisms relevant to present and future threats
4. Creating and institutionalizing infrastructure facilitating sophisticated training and simulation exercises for decision makers
5. Ensuring institutionalized continuity

AIR TRAVEL SECURITY

Since 9/11, the United States, in cooperation with other nations, has sought to increase travel security by creating and emphasizing watch lists, travel guidelines and restrictions, more effective and efficient monitoring of cargo and seeking more sophisticated, less burdensome security checks of airline passengers. Each of these categories constitutes a critical aspect of modern-day international travel security; their successful implementation, with respect to U.S. homeland security, benefits from international cooperation. Needless to say, the Detroit attack decisively demonstrated the inherent vulnerability of these systems.

Watch Lists

After 9/11 the U.S. government created a No Fly List, also known as a watch list, which is a list of individuals who for safety considerations are not permitted to board commercial aircrafts bound for the United States. Currently there are approximately a million people on the government's watch list.[*] Passengers flagged by airport computer systems as false positives (identified as a threat/vulnerability when in fact they are not) are not issued boarding passes until the airline has confirmed the individual is not the *suspected* individual on the watch list. Additionally, false positive passengers endure longer wait times at airports, for they are not permitted to use Internet or kiosk check-in procedures.

On April 28, 2008, DHS announced a new program to remedy watch list mistakes caused by common names. Under the new program, each commercial airline may create a system to verify and securely store a passenger's birth date and other identifying information. DHS will highly regulate each airline's system to ensure that the systems only store identifiable individuals. By voluntarily providing these data and verifying them at a ticket counter, passengers who previously were inconvenienced will now be allowed to board airplanes, hassle-free. Canada has followed the U.S. lead by creating a No Fly List, as part of its Passenger Protection Program, aimed at increasing air travel safety. Canada's No Fly List, which contains between 500 and 2,000 names, combines data from domestic and international foreign intelligence sources.[†]

Travel Guidelines and Restrictions

In addition to watch lists, the United States has bolstered international travel security through travel guidelines and restrictions. For example, on June 2, 2008, DHS introduced the Electronic System for Travel Authorization (ESTA), a new online system that is part of the Visa Waiver Program (VWP). ESTA requires all nationals or citizens of VWP countries who plan to travel to the United States for temporary business or pleasure to obtain electronic travel authorization before boarding a plane or cruise ship. Travelers with valid passports may obtain ESTA electronic authorization as far as two years in advance of their expected travel dates rather

---

[*] http://www.aclu.org/technology-and-liberty/terror-watch-list-counter-million-plus (last accessed October 24, 2010).

[†] http://www.angus-reid.com/polls/1376/canadians_hold_mixed_views_on_no_fly_list/ (last accessed October 24, 2010).

than waiting for authorization (and possible denial) while en route to the United States.

ESTA's paperless environment greatly increases traveler convenience, as travelers no longer need to remember physical forms on busy travel days, and also increases security in the United States by ensuring that all travelers have been authorized prior to travel. Furthermore, ESTA supplants U.S. officials' heavy burden of managing more than 15 million I-94W forms with an immediate, trustworthy online system.

Cargo Transportation and Cargo Security Checks

While watch lists and travel guidelines and restrictions eliminate threats from particular passengers, airplane cargo must also be safely managed; TSA is responsible for ensuring the security of cargo placed aboard passenger airplanes. Following the 9/11 attacks, TSA worked closely with Congress for more than six months to maximize security of air cargo; to that end, TSA both utilizes surprise cargo security inspections called strikes, which results in additional screening of all cargo at a given location, and eliminates all exemptions to screening of air cargo. The majority of TSA's additional background checks focus on cargo employees who screen cargo, cargo employees who have knowledge of how the cargo will be transported, and cargo employees who actually transport the cargo.

In addition to strikes, elimination of exemptions, and increased background checks, TSA has maximized its cargo security by strengthening its cargo security personnel (Figure 4.1). For example, TSA employs 620 cargo transportation security inspectors (TSIs) exclusively dedicated to the oversight of cargo. In addition, TSA has put procedures in place to meet the congressionally mandated requirement of screening 50% of all cargo on passenger planes.*

Despite TSA's attempts to maximize cargo security, many flaws still remain. Most strikingly, TSA maintains a "known shipper" list that lists known shippers TSA does not inspect as stringently as other shippers because the known shippers meet certain regulatory requirements. In fact, unless a package from a known shipper arouses suspicion or is subject to a random search, TSA assumes the package's contents are safe. Perhaps most problematic, U.S. dependence on known shippers lowers other countries' confidence in U.S. cargo security.

---

* http://www.tsa.gov/what_we_do/layers/aircargo/milestone.shtm (last accessed December 20, 2010).

TSA has significantly increased the number of **Transportation Security Inspectors** (Cargo) since FY 2006

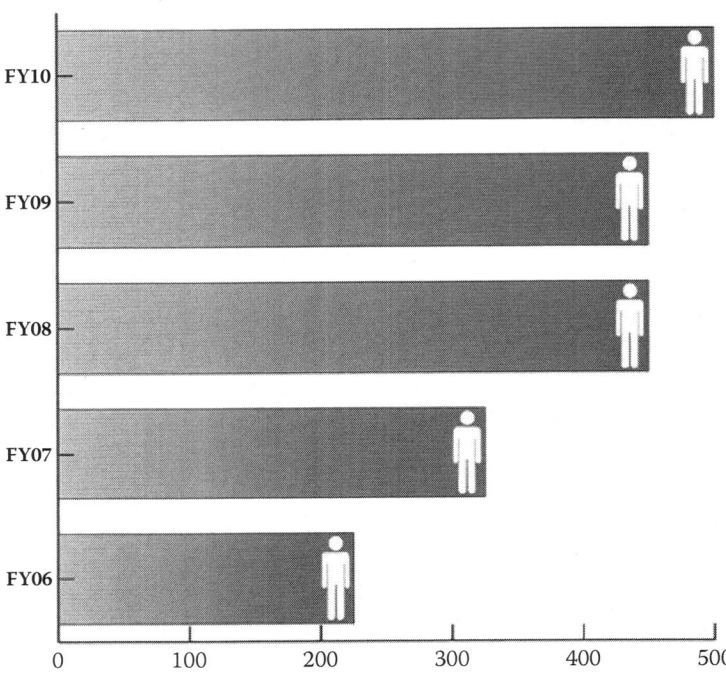

To ensure a high level of compliance with our regulations, TSA employs 620 Cargo Transportation Security Inspectors (TSIs), who are exclusively dedicated to the oversight

FIGURE 4.1   TSI head count since 2006.

IMMIGRATION AND BORDERS

Border Security

The security of U.S. international borders has been an area of great concern post-9/11. The threat of a foreign terrorist entering the United States through porous borders is a very real possibility. To stem the flow of illegal border crossings, the United States has begun the construction of additional barriers along the U.S.-Mexico border, begun testing of a "virtual fence," sent National Guard troops to the U.S.-Mexico border, and hired additional border agents.

The number of illegal immigrants apprehended along the U.S.-Mexico border plummeted by 54% between 2005 and 2009.* In one area where some of the fencing has been built, known as the Yuma sector, there has been a decrease from 800 to 50 individuals apprehended daily in their attempts to cross the border illegally. The question is whether this decrease in arrests is due to government efforts, which have resulted in better enforcement. Critics argue that the decrease in apprehension does not reflect effective border control by the U.S. government, but is rather a result of the current economic crisis in the U.S.

Critics of building physical barriers have argued that illegal border traffic will merely relocate. The virtual fence pilot project is one potential solution to fill in the gaps where physical barriers have not been constructed, consisting of mobile towers equipped with surveillance equipment. Although the program has experienced technological setbacks, it can be credited for the apprehension of 2,000 suspects since September 2007.

Another effective and mobile tool are unmanned aerial vehicles (UAVs) equipped with surveillance instruments. The UAVs also have superior flight times of over 30 hours without having to refuel, allowing for uninterrupted monitoring of the border areas. This also makes UAVs well suited for surveillance of the more remote border areas. Accurate camera images are conveyed from the UAVs in real time to ground operators, who can then have ground agents deployed.

While these measures have seen varying degrees of success, one concern is that most of the border security enhancements have focused on the Mexican border, to the exclusion of the Canadian border. In May 2007 there were nearly 12,000 border agents stationed on the southern border and 972 on the northern border. In contrast, the U.S.-Canadian border spans 5,000 miles, while the southern border covers only 1,900 miles. Although illegal crossings are a bigger problem on the southern border, terrorists may exploit weaknesses in the northern border to gain entry, as was the case with the so-called "millennium bomber."

---

* http://www.foxnews.com/us/2010/06/30/apprehensions-plummet-custom-agents-sky-rocket-border/ (last accessed October 24, 2010).

Passports and Travel

Aside from border infiltration, the other means by which people come to be in the United States illegally is by abusing the legal avenues of entry. This is primarily accomplished in two ways:

1. Overstaying legally obtained visas
2. Using fraudulent travel documents

Visa overstays account for an estimated 30–40% of illegal residents in the United States. In order to combat visa overstay, an entry-exit system that would use biometric data to track visitors has been proposed. However, that system encountered technological and financial problems and has been scaled back, for now, to an entry tracking system only. While this is an improvement, it does little to remedy the problem of not knowing which visitors have outstayed their visas. Recently, the United States implemented changes in travel documents that will likely make falsifying travel documents significantly more difficult. Passports now issued by the United States contain radio frequency identification (RFID) chips that can be used to store biometric data, such as photographs, fingerprints, and iris scans.

The United States also requires any countries that participate in the Visa Waiver Program to issue passports equipped with chips; similarly, the use of biometric identifiers in travel documents helps prevent the possibility of a lost or stolen passport being used by another individual and helps prevent the falsification of travel documents. However, there are concerns about the RFID chips' vulnerability to hacking or other unauthorized access through a radio frequency.

THREAT ASSESSMENT

Maintaining U.S. homeland security inherently involves international cooperation in threat assessment. The United States cannot respond to all threats against the homeland without the appropriate intelligence. What follows is an examination of the U.S. effort to promote intelligence sharing in the context of institutional limits that are—in many ways—the true "bugaboo" of intelligence sharing.

Intelligence Sharing

Intelligence is highly time-sensitive; timely and relevant information directly contributes to effective operational counterterrorism. Timeliness, then, is an essential aspect of intelligence; not only is the information itself important, but *when* it is received is essential to determining its relevance and operability. Therefore, one of the most important concerns in the post-9/11 world has been developing mechanisms to facilitate timely analysis of enormous amounts of information to produce relevant intelligence that can be easily shared and understood by others.

The National Counterterrorism Center (NCTC) represents the main U.S. effort to unite the many intelligence agencies and sources in a way that coordinates and enables the efforts of all the intelligence agencies. Created in 2004, the NCTC consists of a team of intelligence experts from U.S. intelligence agencies and departments. With access to 30 different networks used by the various agencies, NCTC oversees the administration of NCTC online, or NOL. This database allows users the ability to access counterterrorism information and intelligence. Additionally, the NCTC provides "alerts, advisories, warnings, and assessments on topics of interest that are widely disseminated to domestic and overseas operators and analysts."

Opportunities and Concerns

Nevertheless, the involvement of other nations—essential to successful intelligence gathering and analysis—is largely contingent upon their efforts and willingness to utilize available resources and integrate them effectively into their own investigative and intelligence efforts. However, several factors necessarily limit the amount and type of information and intelligence the United States can and should share with other nations. First, security concerns require that sensitive information and intelligence be carefully guarded and made available only to those with a demonstrated need to know.

Clearly, maximizing the availability of intelligence to international partners presents several promising benefits; other nations could integrate U.S. intelligence products with their own in efforts to identify, disrupt, and prevent terrorist attacks and activities. Additionally, finding ways for the United States to share intelligence with other nations will naturally create opportunities to receive and integrate intelligence products from other nations.

That said, it will be necessary to integrate foreign intelligence experts into a national-level intelligence effort in order to capitalize on the resources, knowledge, and understanding of the other nations. While this again raises legitimate security considerations (source protection) regarding introducing foreign actors into intelligence coordination and decision making, any meaningful efforts at international intelligence sharing must go beyond mere sharing of sterilized information.

International Cooperation

The first step to effective international cooperation in homeland security is to forge lasting international partnerships with different countries and multistate organizations, such as the EU. The United States must strive to continue positive trade relations with different countries, as trade is vital to security. In order to promote true, effective partnerships, the United States must maintain open communication with liaisons between partner countries and organizations. Partner countries and organizations should speak through liaisons from counterpart agencies, departments, and organization (such as the Red Cross) from each country or multistate organization, to ensure active communication about security and threat assessment and to promote effective use of international cooperation in counterterrorism.

Coordinated International Security Plan

Partner countries must work together to develop a coordinated international security plan. That plan must outline steps for coordinated travel security, border security, and for determining, and acting on, known and unknown threats. For example, partner countries and multistate organizations must work together to enforce security measures such as watch lists and cargo restrictions. Partner countries must communicate possible threats to travel security, and must create a set plan of action for disaster, terrorist attack, and terror threat scenarios. The plan should outline each multistate organization/country's role within the greater security plan. The plan should articulate the coordination of country-specific agencies, departments, and organizations and outline how each entity must act in the face of a terror threat. A coordinated plan is vital to effective international cooperation in homeland security.

For example, the United States could learn from the EU's cooperative efforts to strengthen its member states' borders. The United States could benefit tremendously from aiding and cooperating both Canada and Mexico in managing their external borders. Increasing the security of our neighbors' borders could help reduce the number of third-country nationals attempting to enter the United States through our northern and southern borders. This is particularly important, as terrorists are likely to attempt to cross those borders in order to avoid the rigors of entering the United States directly from another continent.

International Training

Partner countries must work together to create and undergo international training and simulation procedures. Representatives from partner countries must take steps to undergo specific disaster and terror attack scenario training and simulation to ensure that each member follows the articulated security plan. Further, the training must ensure that each partner country follows the plan within its country, and also that the coordinated plan is followed among all member countries as well. For example, the United States and Canada should undergo border security training and run real-time terrorist border infiltration simulations. Also, the United States and countries like England should undergo travel safety training and simulation of travel security breach scenarios in order to ensure that each participant understands and follows the articulated security plan, and to ultimately ensure effective international cooperation.

Institutionalized Continuity

Finally, the United States and its international partners must ensure institutionalized continuity both between nation leaders and between each nation's key agencies and department liaisons. Institutionalized continuity on an international level refers to the idea that there must be a set process for which to continue, to pass on, the articulated security plan from one nation leader to the next. Each leader, liaison and representative must ensure that he or she understands the coordinated security plan, and must continue to improve upon the technology, intelligence, and training in order not only to develop but also to maintain a high level of international security.

This requires a dialogue between partner nations asking: Does the security strategy work? It requires the creation and continuity of parameters by which to measure international effectiveness. What is the ultimate goal? What are the expectations as to security training and financial ability? Ultimately, there must be an articulated, institutionalized plan to ensure the continuity of security between nation leaders, and all agency/department counterparts, in order to promote effective international cooperation and ensure an effective homeland security strategy.

CONCLUSION

In order to create an effective homeland security strategy, the United States must take measures to promote international cooperation in counterterrorism. After examining the U.S. efforts in travel security, border security, and threat assessment, as well as examining comparative efforts in Canada, Japan, and the EU, it is clear that the United States must follow a two-step process. First, the United States must take measures to protect the homeland. Those measures include promoting travel security by implementing sophisticated technology, promoting border security by securing the northern border, implementing intelligence sharing between agencies, creating a coordinated plan to promote travel and border security, undergoing training and simulation, and finally, ensuring institutionalized continuity from one administration to the next.

After taking action to protect the homeland, the United States must use these factors as a foundation on which to establish international cooperation. To establish effective international cooperation in homeland security, the United States must take measures including the following: forging international partnerships, sharing intelligence related to travel security, creating a coordinated international security plan, running international training and simulation exercises, and finally, implementing international institutionalized continuity.

5

Immigration/Narcoterrorism

INTRODUCTION

Immigration, in the American ethos, is a term filled with inherent con-tradiction. On the one hand, we are a nation of immigrants; on the other hand, American history is replete with political parties and movements seeking to limit immigration. The history of American immigration reflects—or perhaps more accurately influences—the changing nature of demographics. If in the 1800s and 1900s the majority of immigrants were largely from Eastern Europe, Italy, and Ireland, today's immigrant popu-lation is largely from Mexico, the Far East, and South Asia (Figure 5.1).*
In coarse terms, we are a nation of immigrants who seemingly do not like immigrants. The reasons are fairly obvious and predictable: impact on jobs (immigrants will generally accept lower-paying jobs for both skilled and unskilled labor), local communities (schools, language bar-rier issues, public benefits they may receive), and unfortunate, ethnic-based stereotypes.

In addition—in part reflecting inherent suspicion of immigrants (as distinct from immigration)—over the course of history immigrants have been wrongly accused of committing serious crimes with tragic conse-quences. That is, society distinguishes between immigration (generally viewed positively) and immigrants (generally viewed negatively); the

---

* http://ddimick.typepad.com/dennis_dimicks_blog/2008/12/charting-us-immigration-
trends.html (last accessed December 19, 2010).

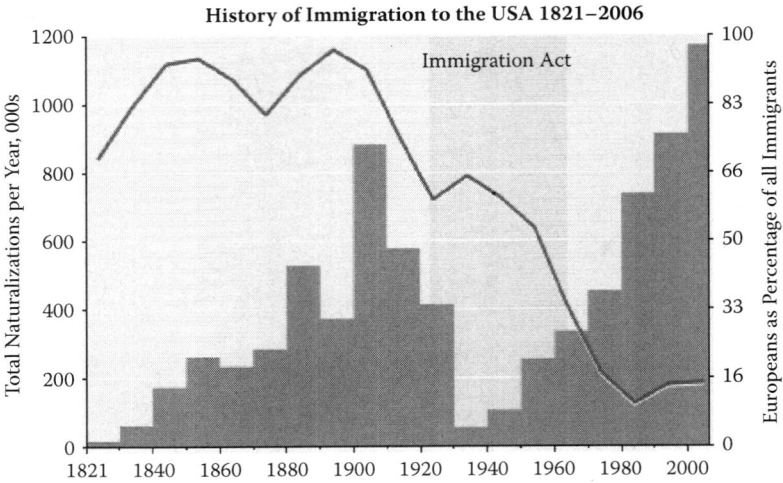

FIGURE 5.1 A timeline of the origins and numbers of immigrants to the United States.

concept of immigration is positive, while its practical implementation and reality is viewed otherwise.

As will be discussed in this chapter, a confluence of circumstances and conditions has served to reignite the immigration and, by extension, immigrant discussion, pushing both issues to the forefront of the contemporary political debate. The dialogue has been marked by strident debate, court action, and intensive public scrutiny, controversial political decisions. Simply put, the tone and tenor of the discussion surrounding, in particular, Arizona Senate Bill 1070 is but a manifestation—perhaps rearticulation—of an age-old discussion regarding the status, rights, and "place" of immigrants and aliens. Needless to say, it is not the first instance in which the "other" has been singled out, whether justifiably or not. The underlying rationalization, both in the current milieu and historically, for singling out immigrants reflects core prejudices and deep fears. In the post-9/11 world of super-heightened tensions and anxieties highlighted by increasing concern regarding domestic terrorism, and therefore homeland security vulnerabilities, the immigrant question takes on new, powerful meanings.

However, the question—fear aside—is whether immigrants, regardless of their status, pose a threat to American national security. That question cuts to the core of whether homeland security is viewed through the lens of person-specific or group-specific threat analysis. Rearticulated: Does law enforcement, in the context of homeland security, focus on a particular individual predicated on intelligence information subject to rigorous and robust review, or is the assumption that individuals belonging to distinct ethnic and racial groups pose a threat, regardless of specific information regarding a particular individual? This is the obvious danger of identifying immigrants (and immigration) as a threat rather than engaging in the unquestionably more complicated law enforcement effort requiring person-specific intelligence gathering/analysis.

There is, as the dark pages of history clearly manifest, a disturbing tendency to view immigrants as posing a threat. Essential to our discussion and a guide to the pages ahead is whether the identification of immigrants as a threat is grounded in reality or based on prejudice, whether it is predicated on concrete person-specific information or whether the basis is a dangerous approach best defined as "them."

It is essential to recall that immigrants are *highly* vulnerable members of society seeking to make a new life for themselves and their families; they are the contemporary embodiment of the American dream. That definition applies to those in the United States *both* legally and illegally; those illegally in the United States are no less an embodiment of the American dream than those legally living in the United States. That is to neither excuse nor justify the actions of wrongdoers; it is, however, to highlight the existential reality of immigrants: a desire to create a better future for their children regardless of their status. However, as demonstrated throughout American history, the body politic—themselves children of immigrants—does not necessarily respond with open arms.

President Obama, as a candidate, promised immigration reform, whereby illegal immigrants would become legal. That is, the legislation would facilitate status change for individuals who had entered the United States devoid of requisite documentation. While obviously controversial, Obama's promised efforts would have both practical and philosophical significance, the former, because it would remove the constant threat of arrest and deportation for those illegally in the United States, the latter, because it would suggest an acceptance of immigrants, not only of immigration.

While Obama's discussed but not submitted (as of this book's publication) legislative proposal reflects a particular political paradigm, it is obviously not one universally shared. The fundamental concern raised with respect to the discussed proposal focuses on a perception that

71

immigration reform is akin to "blanket amnesty" for those who entered the United States illegally.* In other words, the immigration reform debate highlights—starkly and clearly—a significant contradiction in American politics: on the one hand, it manifests the American ideal of new beginnings and promises, and on the other hand, it manifests America's inherent fear of the immigrant and his impact on the economy, especially those marginally employed. In other words, there is a huge *ongoing* tension between liberal principles and xenophobia. This irresolvable dichotomy has been at the forefront in the failure—throughout American history—to satisfactorily resolve the immigrant question. It is, in many ways, an ongoing, seemingly irresolvable dilemma that today has become a powerful—and perhaps dangerous—political football. Reflecting contemporary political realities, the Obama administration is backtracking on promises regarding immigration reform.†

The relationship between this paradox and homeland security is clear; in response to a threat (perceived or actual) on America, the "outsider" is the easy, most vulnerable target. While that is inherently dangerous to both the immigrant's physical safety and society's values, it is a reality; examples abound worldwide, particularly in response to either economic downturn or racial/ethnic tensions. Riots in France by angry immigrants in response to alleged systematic and continuous police brutality against immigrants, repeated attacks against immigrants in Germany,‡ and growing anti-Islam sentiment in the Netherlands as articulated by Gert Wilders and as manifested in the 2010 parliamentary elections are but the most obvious examples of anti-immigrant sentiment in Europe.§ The present wave of anti-immigrant sentiment in certain parts of Europe reflects an increasing concern regarding the growth of Islamic communities. That concern has called into question whether the European model of multiculturalism has not resulted in an "own goal" by facilitating the development of parallel societies whereby immigrant communities live beyond mainstream society.

---

* http://azstarnet.com/news/local/article_83ea98f0-4433-11df-aec7-001cc4c002e0.html (last accessed April 10, 2010).

† http://www.azcentral.com/arizonarepublic/news/articles/2010/04/06/20100406immig-momentum0405.html (last accessed April 10, 2010).

‡ http://www.nytimes.com/2000/08/01/world/german-faults-silence-about-attacks-on-immigrants.html (last accessed December 20, 2010).

§ http://www.cnsnews.com/news/article/60884 (last accessed December 20, 2010).

The potential dangers from distinct, parallel societies are clear, as highlighted by the Madrid train bombing in 2004,* the London tube bombing,† the Glasgow Airport attack,‡ the alleged (but foiled) simultaneous attacks on 13 commercial airlines scheduled to fly from London to the United States, and the brutal killing of Theo van Gogh.§ In each of these attacks the attackers—Islamic extremists—were either citizens or legal immigrants living and working among their fellow countrymen. However, in retrospect, those responsible for these attacks—predicated on religious extremism—were disenfranchised from mainstream society even though (particularly in the UK) their socioeconomic status was best defined as middle class/upper middle class and they had seemingly, successfully, integrated into British culture and society. While this is not to suggest that European trends automatically translate into American patterns, recent attacks in the U.S. suggest similar trends.

This topic—future threats and patterns, the "future face of terrorism"—will be discussed in depth in Chapter 6; its relevance to the immigrant discussion is the threat (direct or indirect) posed to society by a group or individuals perceived as outsiders, either by larger society or predicated on self-perception. In both cultures—European and U.S.—the question is whether today's outsider wants to become tomorrow's insider; simply put, does he or she want to acculturate and assimilate or disassociate from the nation-state? While in Europe there is increasing concern regarding the possible threat posed by the parallel society paradigm, in the United States, the *traditional* model has been integration in the work force but self-contained communal existence (i.e., Little Italy, Chinatown, etc.), but with an embrace of the American dream and the opportunities it presents. As we move forward, one of the important issues to explore is whether that traditional model is still, universally, in play.

In that vein, the threat *potentially* posed by the outsider is the dominant aspect of resistance to immigration reform, much less to actual immigration. From the general to the particular, in the context of homeland

---

* http://news.bbc.co.uk/2/hi/in_depth/europe/2004/madrid_train_attacks/ (last accessed December 20, 2010).
† http://news.bbc.co.uk/2/hi/in_depth/uk/2005/london_explosions/default.stm (last accessed December 20, 2010).
‡ http://www.timesonline.co.uk/tol/news/uk/article2009765.ece (last accessed December 20, 2010).
§ http://www.wsws.org/articles/2004/nov2004/gogh-n10.shtml (last accessed December 20, 2010).

security post 9/11, the question is whether immigration presents a legitimate threat to American national security. The threat is divided into two distinct categories: suspicion regarding members of the Islamic community and suspicion regarding Mexicans coming into the United States, purportedly illegally. Why these two communities in particular? Perception is critical to the discussion. In some ways, in an emotion-laden subject—which both homeland security and immigration are—perception may be more valid and relevant than reality.

IMMIGRANT COMMUNITIES— POSSIBLE RADICALIZATION

In addressing contemporary immigrant communities the point of departure is whether the traditional acceptance—if not embrace—of the American dream by immigrants is being rejected by self-radicalized immigrant groups. After all, in the immediate recent past more than 20 young Somali-Americans[*] living in Minneapolis have disappeared from the United States; according to the FBI, "some of them went to Somalia to fight with the Islamic extremist group, al-Shabab."[†]

Of particular concern, the FBI believes that one of the members of the group, Shirwa Ahmed, "was the prime suspect in an October 2008 suicide bombing in Somalia,"[‡] making him the first known U.S. citizen suicide bomber. According to officials, one of the principal concerns regarding the "lost boys" is "that some of the men may be destined to return to the US after they receive terrorist training."[§] The term has been ascribed to men in their twenties who literally disappear from their homes and shortly thereafter telephone their families

---

[*] http://www.nytimes.com/2009/07/12/us/12somalis.html?_r=2&hp.

[†] http://www.voanews.com/english/archive/2009-03/2009-03-27-voa43.cfm?CFID=2561 29143&CFTOKEN=57091416&jsessionid=663015a2bd228a5e3567537e66d163c201d3 (last accessed October 31, 2009).

[‡] http://www.voanews.com/english/archive/2009-03/2009-03-27-voa43.cfm?CFID=2561 29143&CFTOKEN=57091416&jsessionid=663015a2bd228a5e3567537e66d163c201d (last accessed October 31, 2009). At least 30 people were killed in this attack; http://www.star-tribune.com/local/40202352.html?elr=KArksUUUU (last accessed October 31, 2009).

[§] http://www.defenddemocracy.org/index.php?option=com_content&task=view&id=1178 3918&Itemid=347 (last accessed October 31, 2009).

from Somalia providing little additional information.* What has been referred to as the Somali-Minneapolis Terrorist Axis† is facilitated by a combination of social isolationism (in schools) and radicalization (outside the home). Furthermore, and perhaps more importantly, family members‡ blame the Abubakar As-Saddique Islamic Center in Minneapolis for the boys' decisions to travel to Somalia and join militant Islamic groups.

The FBI has focused its efforts on the center since it provides the common link between the boys.§ As FBI Director Robert Mueller succinctly stated, "It appears that this individual [reference is to the suicide bomber, Ahmed] was radicalized in his hometown in Minnesota."¶ Mueller's assessment has led the FBI to engage in surveillance of the mosque, a measure that while criticized by CAIR, has been met with support by some family members of the 'lost boys.'"**

Examining immigrant communities and multiculturalism leads to one fundamental question: What is the relationship between the immigrant community and the host country? In essence, if the members of the immigrant community live in a parallel society, segregated from mainstream culture, rather than functioning as vibrant contributing members of the host country, red flags regarding multiculturalism's beneficence must be raised. Brian Berry has suggested that while assimilation requires ratification by the receiving group, in acculturation the individual comes to acquire cultural practices belonging to a tradition of another group.††

Parallel societies, or what Tariq Modood calls "creating an alternative society,"‡‡ pose a significant danger to liberal society because, as Modood explains, they foster or shelter radicalism. Disturbingly,

* http://i.abcnews.com/US/WireStory?id=6344443&page=2 (last accessed October 31, 2009). One individual, Burhan Hassan, was killed in Somalia. According to his family, he had decided to return to the United States but was killed because he could identify who had recruited him; http://www.npr.org/templates/story/story.php?storyId=105572589&ft=1&f=1001 (last accessed October 31, 2009).
† http://www.npr.org/templates/story/story.php?storyId=105572589&ft=1&f=1001 (last accessed October 31, 2009).
‡ http://www.msnbc.msn.com/id/29620604/ (last accessed October 31, 2009).
§ http://wcco.com/local/mosque.leaders.no.2.996582.html (last accessed October 31, 2009).
¶ http://islaminaction08.blogspot.com/2009/03/fbi-watching-somalian-muslims-in.html (last accessed October 31, 2009).
** http://www.foxnews.com/wires/2009Mar10/0,4670,TwinCitiesSomalis,00.html (last accessed October 31, 2009).
†† Brian Barry, Culture and Equality (Harvard University Press, 2001), 73.
‡‡ Tariq Modood, Multiculturalism (Polity Press, 2007).

radicalism manifests itself in the immigrant community in two primary ways: sexual repression and political violence. The inherent isolationism of segregated communities makes the state largely unable—perhaps unwilling is a more accurate term—to engage those that it otherwise would. Political philosophers argue that the essence of liberal society is tolerance of diverse communities when the state encourages individual expressions of speech and conduct. However, according to Martha Minow,* how much intolerance can be tolerated is a pressing question demanding our fullest attention. Minow's suggestion is particularly poignant in the context of immigrant communities whose illiberalism—predicated on the mores of their "former" cultures—runs counter to the liberal societies that nevertheless tolerate them even though harm occurs to internal apostate members.

LEGAL AND ILLEGAL IMMIGRANTS

A discussion regarding immigrants is complicated by the legal and political reality that the community must be divided into two distinct categories: legal and illegal.† In the last decade, 13.1 million legal immigrants settled in the United States; although difficult to discern the exact number of illegal immigrants presently in the United States, estimates range between 12 and 20 million.‡ The political discussion, with respect to the immigrant question, has largely focused on illegal immigrants and what should be their status—if any—given that they both entered and reside in the United States illegally. This dilemma—at the heart of significant political controversy—has assumed greater importance in the contemporary homeland security discussion.

The reason for this is twofold: while 9/11 drew much attention to the rights, protections, guarantees, and privileges extended to non-Americans *legally* living in the United States, narcoterrorism and the extraordinary violence in Mexico near the U.S.-Mexican border has greatly heightened concerns regarding the powerful and dangerous convergence between drugs, violence, and *illegal* immigrants from Mexico. The three—separately and

---

* Martha Minow, Tolerance in an Age of Terror, 16 S. Cal. Interdisc. L.J. 453 (2007).
† http://www.usimmigrationsupport.org/amnesty.html (last accessed April 10, 2010).
‡ http://www.usimmigrationsupport.org/illegal-immigration.html (last accessed December 20, 2010).

collectively—are neither necessarily nor instinctually homeland security issues. However, given their short-term and long-term homeland security implications and applications, it is necessary to include drugs and violence under the umbrella of homeland security.

That is not to suggest that either is not relevant to other paradigms, in particular law enforcement. It is, however, to suggest that viewing both broadly—in accordance with the 18 critical infrastructures tasked to DHS—compellingly and decisively suggests that immigration (and immigrants) and the threats it potentially poses is, indeed, a homeland security issue. While, perhaps, not exclusively a homeland security concern, the threats above are increasingly identifiable as such. The caveat, in the context of constitutional law, is that the identified threat must be person specific, not based on a "round up the usual suspects" approach suggestive of "guilt by association."

Discussing immigration, in the context of homeland security, requires distinguishing between legal and illegal immigrants. However, it does not tell the tale in its entirety because those responsible for 9/11 were in the United States legally. That is, to suggest that illegal immigration is the heart and soul of the issue is misleading. Perhaps it scores political points in certain forums; however, it is an insufficient—and politically deceptive—analysis precisely because those who committed 9/11 entered and resided legally in the United States. Questions regarding criteria for visa issuance are legitimate and have come under increasing scrutiny in the aftermath of 9/11; nevertheless, the illegal immigrant discussion does not apply to those responsible for 9/11.

The legality of their presence, after all, resulted from U.S. consular officials granting visas facilitating lawful entry into the United States on the assumption that the individuals would not pose a threat to homeland security (9/11 is an obvious exception), whereas illegal immigrants are increasingly viewed as posing a threat. Needless to say, that is not to suggest that all illegal immigrants fall into this category; it is, however, to highlight the distinct—*perceived*—contemporary threats posed by individuals entering the country illegally. While historically illegal immigrants were either job seekers whose visa application was denied or individuals who never submitted applications, assuming rejection, the present political and social reality is that illegal immigration is viewed as a significant homeland security issue precisely because of its attendant dangers and threats.

DRUG SMUGGLING AND IMMIGRATION

In many ways, then, the immigration discussion requires circling back to an issue previously analyzed: What is homeland security and what threats (perceived and actual) are posed by immigration, legal and illegal alike? Immigration—whether one favors reform or not—cuts across many questions related to the American ethos. After all, five years after moving to America (provided the entrance to the United States was legal)—dependent on meeting various statutory regulations—the immigrant is entitled to become an American citizen with all its rights, obligations, and responsibilities. That said, the concern regarding the danger, regardless of its palpable political incorrectness, must be addressed, for it is a critical contemporary reality. Simply put: Do immigrants pose a threat to America?

Let us begin with immigrants seeking to come north from Mexico. Is this the manifestation of the American dream or encumbering America (and Americans) with significant social tension and conflict indigenous to Mexican society, particularly with respect to the dramatic rise in drug-related crime and corruption? Of particular concern to the American public is whether the drug wars (and gangs) will be exported to the United States, in particular the Southwest.

This raises an important—and legitimate—question: Does the possibility that Mexican drug-related gangs will penetrate certain communities in the United States present a criminal or homeland security threat to the American public? While the instinctual response would suggest this is exclusively a criminal matter as the convergence between gangs and drugs has traditionally been perceived in the American model there is something fundamentally different with respect to the present danger. While gangs and drugs are not new to the American scene, the physical immediacy of an international-based threat suggests this is not akin to former drug-related crimes nor the traditional criminal law paradigm.[*]

The discussion whether—and to what extent—illegal immigrants pose a threat to U.S. homeland security is laden with political overtones and "spin." In order to determine the validity of the threat(s), the opinions of a number of experts were solicited. While these individuals were exceptionally candid with their comments and analysis, the condition for their cooperation was a guarantee of anonymity. Given the importance, relevance, and validity of their comments, I have decided to include them,

---

[*] http://www.latimes.com/news/opinion/commentary/la-oe-poizner27-2010mar27,0,5816792. story (last accessed December 20, 2010).

78

both because of a determination on my part their comments are agenda-free and because of important light they shed on a most complicated and contentious subject. My conversations with these individuals focused, in large part, on the possible confluence between terrorism and illegal immigrants from Mexico, with an obvious emphasis on drug gangs and the smuggling of drugs and people into the United States.

As background to this section, when serving as the Israel Defense Force's legal advisor in the Gaza Strip, I heard it forcefully argued that a direct link exists between terrorism and illegal drugs. That link, in particular, focused on travel routes used by both terrorists and drug couriers. In posing the question whether terrorists could use the same routes by drug gangs coming from Mexico into the United States, the consensus response (with wiggle room) was that the Israeli model was not relevant to the Mexican paradigm. That said, on March 9, 2010, the chief of the U.S. National Guard Bureau, General Craig McKinley, warned of the growing threat to U.S. national security through "the linkages between drug cartels through organized crime back to terrorist organizations [which] cannot be disputed."[*]

In large part the assessment of the individuals with whom I spoke—in contrast to General McKinley's assessment—was based on a conviction that Mexican drug gangs are not interested in sharing routes or information with terrorists, as that would potentially work against their self-interest. With respect to illegal immigrants coming from Mexico, the increasing difficulty faced by those seeking to enter the United States has directly led to a dramatic increase in fees demanded by smugglers; if in years gone by smugglers demanded $50 for smuggling an individual, today's smugglers demand $3,000.

To that end, the smuggling industry is criminalized, sitting at a combustible confluence between human and drug smuggling as the know-how (routes, terrain, climate, location of U.S. border officials) is strikingly similar. Furthermore, the ability of the smugglers (drugs and human alike) is facilitated by the corruption of local Mexican officials in the states bordering the United States and, arguably, the limited ability of U.S. agents.[†] Furthermore, the smugglers are increasingly sophisticated and institutionalized, developing and implementing business models significantly

---

[*] http://www.metimes.com/security/2009/03/09/general_drug_cartels_linked_to_terrorism/0009/ (last accessed December 20, 2010).

[†] http://www.fsrn.org/audio/report-finds-problems-with-police-role-immigration-enforcement/6495 (last accessed December 20, 2010).

facilitating their efforts. The business model—largely predicated on the profitability of smuggling (drugs and people alike)—is akin to decentralized enterprises with extraordinary local autonomy rather than an overarching organization with broad powers and reach.

That said, as has been consistently demonstrated in Afghanistan, drug couriers use travel routes and methods used by terrorists.* The connection between drugs and terrorism is powerful enough to suggest that the former must not be viewed exclusively as a criminal matter. Simply put, if drugs and terrorism are linked—whether directly or indirectly—then the threat to homeland security is more relevant than otherwise believed. Perhaps the link would have been considered untenable in years past, but an expansive view and application of homeland security suggests an intertwining not previously considered. This is particularly—and increasingly—relevant in the context of courier routes.

9/11 AND IMMIGRANTS

Because those responsible for the 9/11 attacks were in the United States legally, decision makers and the public alike were forced to revisit the issue of legal immigrants. Even though the terrorists responsible for 9/11 had not violated American law the public demanded to know whether entry requirements were sufficiently restrictive or unnecessarily lenient. If, indeed, entry requirements were unduly lax, thereby facilitating lawful immigration to America by individuals posing a threat to national security, then rearticulation of standards was justified. The question is neither ephemeral nor abstract; rather, it is direct and concrete. If the standards, guidelines, and criteria enabling immigration to the United States are broad and vague, then—so goes the argument—there is inadequate threat analysis regarding possible sources of danger. The inevitable panic response in the immediate aftermath of 9/11 was detention of individuals deemed to pose a threat.

How was that potential threat defined? In reality, as discussed in the section below, the threat was extraordinarily broadly defined, divorced from rigorous person-specific analysis. In essence, the administration targeted a category rather than individuals. While perhaps the direct result of a massive intelligence failure, what is significant, in the context of

---

* http://www.washingtontimes.com/news/2009/mar/27/hezbollah-uses-mexican-drug-routes-into-us/ (last accessed April 10, 2010).

homeland security is the initial response. That response was to identify immigrants—those "not of us"—as posing direct and immediate threats. Admittedly immigrants were involved in 9/11; however, not all immigrants were involved in 9/11. The instinctual reaction to lump all immigrants into a category best defined as "posing a danger" has a twofold impact: while perhaps satisfying public clamor for action regardless of effectiveness, it unnecessarily burdens law enforcement with unwarranted demands, thereby all but guaranteeing ineffectiveness. Perhaps identifying immigrants as a category makes for good political theater; however, it makes for ineffective homeland security, as it is devoid of risk analysis, prioritization, resource allocation, and cost–benefit analysis.

Immigrants are a particularly vulnerable class in society, viewed skeptically, if not suspiciously, by the larger community, yet expected to conduct themselves in full accordance with the laws of the land. While societal skepticism is perhaps understandable from a social and cultural perspective, that does not justify suggesting that immigrants—as a class—are a danger to society. Quite the opposite: American history is replete with examples of extraordinary contributions made by immigrants to their new society. After all, the American dream is the intellectual and philosophical articulation of an immigrant society. To that end, the stirring words on the Statue of Liberty greeting millions who came to the shores of America seeking to create a better future for their children articulate an extraordinary ideal:

> Give me your tired, your poor,
> Your huddled masses yearning to breathe free,
> The wretched refuse of your teeming shore.
> Send these, the homeless, tempest-tossed to me.
> I lift my lamp beside the golden door.

Nevertheless, reality has not always equaled the inscription written by Lazarus; herein lies the fundamental tension between the articulated ideal of immigration and the reality faced by immigrants.

ETHNIC PROFILING

In the weeks after September 11, Gallup polls showed almost 60% of Americans favored "requiring people of Arab descent to undergo special, more intensive security checks when flying on American planes," and half said that "they should have to carry special identification cards with them at all times."[*]

> Before September 11, about 80 percent of the American public considered racial profiling wrong. State legislatures, local police departments, and the President had all ordered data collection on the racial patterns of stops and searches. The U.S. Customs Service, sued for racial profiling, had instituted measures to counter racial and ethnic profiling at the borders. And a federal law on racial profiling seemed likely.
>
> After September 11, however, polls reported that 60 percent of the American public favored ethnic profiling, at least as long as it was directed at Arabs and Muslims. The fact that the perpetrators of the September 11 attack were all Arab men, and that the attack appears to have been orchestrated by al Qaeda, led many to believe that it is only common sense to pay closer attention to Arab-looking men boarding airplanes and elsewhere. And the high stakes—there is reason to believe that we will be subjected to further terrorist attacks—make the case for engaging in profiling stronger here than in routine drug interdiction stops on highways....
>
> Press accounts made clear that whether as a matter of official policy or not, law enforcement officials were paying closer attention to those who appear to be Arabs and Muslims. And in November, the Justice Department announced its intention to interview 5,000 young immigrant men, based solely on their age, immigrant status, and the country from which they came.[†]

Historically, racial profiling has not worked, regardless of whether used in the criminal paradigm or in the fight against terrorism.[‡] David Harris, a professor of law and values, argues that when race or ethnic appearance is used in law enforcement, the accuracy of catching criminals decreases.[§] As seen in London in the aftermath of the train bombings of the summer of 2005, the use of profiling can also lead to accidental deaths.[¶] The New York City Police Department initiated a stop and frisk campaign in the late 1990s whereby police would regularly stop people on the street in order to

---

[*] Maia Davis, A Painful Reminder for Japanese-Americans, The Record (Bergen County, NJ), September 20, 2001, a15.

[†] David Cole and James X. Dempsey, Terrorism and the Constitution 168 (New York: The New Press, 2002).

[‡] Kim Zetter, Why Racial Profiling Doesn't Work, Salon.com, August 22, 2005, http://www.salon.com/news/feature/2005/08/22/racial_profiling/.

[§] David Harris, Profiles in Injustice: Why Racial Profiling Cannot Work (The New Press, 2002).

[¶] London police fatally shot an innocent Brazilian man suspected of involvement in the bombings.

confiscate illegal weapons and reduce crime.* Minority communities felt they were unfairly targeted, which turned out to be correct.† Only after Amadou Diallo, an unarmed West African immigrant, was killed during such a stop was it demonstrated, through a study ordered by then New York attorney general Eliot Spitzer, that minorities were unfairly targeted.‡

However, the federal government did not learn from the experience of New York State. In November 2001, Attorney General Ashcroft announced that the FBI and other law enforcement would interview more than 5,000 men mostly from the Middle East, all in the United States on temporary visas.§ However, of the 5,000 individuals initially designated, only about half of these individuals could be found for interviews.¶ Additionally, very little information was learned from those that the FBI was able to locate.** In fact, this tactic has been greatly criticized by former FBI agents, claiming that it is ineffective and "guts the values of our society."††

The ultimate test in balancing civil liberties and freedoms is to determine whether a policy that negatively impacts a particular population group also contributes to a nation's security. If the policy does not have a proven, positive effect, then not only has the desired balance not been struck, but the policy is simultaneously ineffective, problematic, and potentially unconstitutional. A policy that suggests a "lashing out" rather than a calculated response will not—under any circumstance—be considered balanced.

In the context of homeland security, the question to be addressed is whether the administration's actions contributed—directly or indirectly—to preventing additional attacks. If it did, then the additional query at what cost and whether applied measures met constitutional standards of due process and equal protection must, similarly, be examined.

> The government has detained over 1200 persons in connection with its investigation of the attacks of September 11, yet as of late December only one had been charged with any involvement in the crimes under

---

* Id.

† Id.

‡ Id.

§ Jena Heath, Bush Defends Anti-Terror Fight, Military Courts, Atlanta Journal and Constitution, November 30, 2001, A17.

¶ Michael Ratner, Making Us Less Free: War on Terrorism or War on Liberty? http://www.ccr-ny.org/v2/viewpoints/viewpoint.asp?ObjID=YLhsqUx1eu&Content=143 (last accessed March 21, 2006).

** Id.

†† Jim McGee, Ex-FBI Officials Criticize Tactics on Terrorism, Washington Post, November 28, 2001, A1.

investigation, and the government claims that only ten or twelve of the detained are members of al Qaeda, the organization said to be responsible for the attacks. The vast majority are being held on routine immigration charges under unprecedented secrecy. The government will not disclose most of their names, their trials are held in secret, and their cases are not listed on any public docket.

At the same time, ethnic profiling is being broadly engaged in, and widely defended as reasonable. Already, we fear, the government has overreacted in a time of fear, assuming powers in the name of fighting terrorism that are in no way limited to counterterrorist investigations. It has not shown that the new powers it has asserted are necessary to fight terrorism. And it has targeted the lion's share of its infringements on liberty at immigrants, and particularly Arab and Muslim immigrants.[*]

The cost is on different levels. Racial profiling implies guilt by association, a concept abhorrent to American democratic values and one that raises serious constitutional questions.[†] Guilt by association suggests that the actions of the individual are not significant; rather, belonging to a particular religious, ethnic, or social group is sufficient cause to determine guilt. In the context of counterterrorism, it is unclear how such a policy contributes to the nation's security. As demonstrated by Israel's experiences, potential terrorists do not always fit a preconceived notion of appearance or ethnicity.[‡] While there are some who argue that the effectiveness of racial profiling is unclear,[§] its constitutionality is very much in doubt.

---

[*] David Cole and James X. Dempsey, Terrorism and the Constitution 149 (New York: The New Press, 2002).

[†] NAACP v. Claiborne Hardware Co., 458 U.S. 886, 920, 102 S.Ct. 3409, 3429 (1982). (The court held that "for liability to be imposed by reason of association alone, it is necessary to establish that the group itself possessed unlawful goals and that the individual held a specific intent to further those illegal aims." The court also held that "the State may not employ 'means that broadly stifle fundamental personal liberties when the end can be more narrowly achieved.'" Citing Shelton v. Tucker, 364 U.S. 479, 488, 81 S.Ct. 247, 252, 5 L.Ed.2d 231 (1960) and Carroll v. Princess Anne, 393 U.S. 175, 183–184, 89 S.Ct. 347, 353, 21 L.Ed.2d 325.). See also Scales v. U.S., 367 U.S. 203 (1961); U.S. v. Robel, 389 U.S. 258, 262 (1967); and Keyishian v. Board of Regents, 385 U.S. 589, 606 (1967) as examples where the Supreme Court has upheld the right to associate with a particular group, in these cases the Communist Party.

[‡] See Kim Zetter, Why Racial Profiling Doesn't Work, Salon.com, August 22, 2005. (In 1986, a pregnant chambermaid from Dublin was found with a bomb in her carry-on; three Japanese passengers took guns from their checked luggage and opened fire at Ben Gurion International Airport in 1972; in 1970 a Nicaraguan tried to hijack an Israeli El Al plane.)

[§] Sherry F. Colb, The New Face of Racial Profiling: How Terrorism Affects the Debate, FindLaw's Writ, October 10, 2001, available at http://writ.findlaw.com/colb/20011010.html (last accessed March 21, 2006).

American history is fraught with overreactions in times of crisis; in developing a national counterterrorism strategy, the question is whether such methods are effective. The discussion, however, is not limited only to matters of policy. There are serious questions concerning the constitutionality of such policies; identification or designation of aliens as a unique category of individuals *inherently* endangering America suggests guilt by association.

The Supreme Court on a number of occasions has expressly stated that guilt by association is unconstitutional.[*] In *NAACP v. Claiborne Hardware Co.*, the court held that the First Amendment "restricts the ability of the State to impose liability on an individual solely because of his association with another."[†] In holding that petitioners, who demonstrated peacefully, could not be held liable for the damages caused by those who were violent, the court wrote that to "punish association with such a group, there must be 'clear proof that a defendant specifically intends to accomplish the aims of the organization by resort to violence.'"[‡] However, the Supreme Court has upheld the convictions of individuals for membership in a group that advocates the overthrowing of the U.S. government by force or violence.[§] In order for membership to be illegal though, the court held that the membership must be knowing, active, and purposive as to the organization's criminal ends.[¶]

Does this seeming imbalance contribute to a safer America? The balance equation cannot be measured or studied in a vacuum. Government's responsibility is to guarantee the safety of innocent individuals living in America and their property. Simultaneously, the government must uphold constitutional guarantees to those suspected of committing a crime. The ultimate question is whether guilt by association, which is unconstitutional, has been effective. While the U.S. government recognizes that racial profiling, in and of itself, is wrong, the Justice Department maintains that "the racial profiling guidance recognizes that race and ethnicity may be used in terrorist identification, but only to the extent permitted by the nation's laws and the Constitution."[**]

---

[*] See, e.g., NAACP v. Claiborne Hardware Co., 458 U.S. 886 (1982); U.S. v. Robel, 389 U.S. 258 (1967); and Keyishian v. Board of Regents, 385 U.S. 589 (1967).

[†] Claiborne, supra note 33 at 918–919.

[‡] Id. at 919 citing Scales v. U.S., 367 U.S. 203, 229 (1961) and Noto v. U.S., 367 U.S. 290, 299 (1961).

[§] Scales v. U.S., 367 U.S. 203 (1961).

[¶] Id. at 209.

[**] Fact Sheet Racial Profiling, Department of Justice, June 17, 2003 (on file with author).

Therefore, in the homeland security debate, assuming the administration's actions were intended to contribute to preventing additional attacks—as was the case in the operation to capture/kill bin Laden in Pakistan—the attendant social and legal costs of the measure must be addressed to determine what has been achieved. In the case of bin Laden, already strained relations and mistrust between the U.S. and Pakistan grew in the immediate aftermath of the operation. The legality of the operation was questioned as was Pakistan's loyalties to the U.S. as an ally in the "war on terror." It was a calculated risk from a number of angles. At home, targeting immigrants as immigrants, rather than identifying individuals posing specific threats, is the manifestation of ineffective homeland security policy.

In developing a homeland security strategy, the question is whether strategies such as racial profiling are effective. The discussion, however, is not limited just to matters of policy; the costs of this policy must also be addressed. The cost of racial profiling, as a national security strategy, on civil liberties arises at different levels. "Racial profiling implies 'guilt by association,' a concept both abhorrent to American democratic values and one that raises serious Constitutional questions. Guilt by association suggests that the actions of the individual are not significant; rather membership in a particular religious, ethnic, or social group is sufficient to determine guilt. In the context of counterterrorism, it is unclear how such a policy contributes to the nation's security."*

Over the years, the Supreme Court has ruled on issues related to aliens. In *Yick Wo v. Hopkins*, the Supreme Court held that discrimination based on race and nationality was illegal, and that "all persons within the jurisdiction of the United States shall have the same right in every State and Territory ... to the full and equal benefit of all laws and proceedings for the security of persons and property as is enjoyed by white citizens."† Within 60 years though, the Supreme Court allowed these same rights to be taken away. In *Korematsu*, the Supreme Court upheld President Roosevelt's decision to place innocent American citizens of Japanese ancestry in internment camps following the Japanese attack on Pearl Harbor.‡

---

* Id.
† Yick Wo v. Hopkins, 118 U.S. 356, 369 (1882).
‡ Korematsu v. U.S., 323 U.S. 214, 65 S. Ct. 193 (1944). The number of people interned ranges from 110,000 to 120,000. See San Francisco Museum website. Executive Order 9066 requiring the internment of Americans of Japanese heritage was signed on February 19, 1942. The order was rescinded by President Franklin D. Roosevelt in 1944, and by the end of 1945 the last internment camp was closed. http://www.infoplease.com/spot/internment1.html.

Though the United States is a nation of immigrants who traveled far and wide to reach America's shores in an attempt to achieve the American dream, American history is replete with examples of their mistreatment. Employment ads and help wanted signs in many cities had notations saying "No Irish Need Apply,'" and several states refused to allow Jewish immigrants to vote until the mid-1800s.[†] Furthermore, secret societies such as the Know Nothing Movement were formed to restrict non-Anglo Saxon immigration.[‡] In the Palmer Raids of World War I, "suspected radicals were rounded up, held without bail and often deported without trial."[§] These "radicals" were often Eastern European Jews,[¶] and as a result, more than 3,000 aliens were deported.[**] Professor David Cole has compellingly written about the U.S. history of targeting immigrants during a time of fear. During World War I, dissidents were imprisoned for "merely speaking out against the war," and most were immigrants.[††] During World War II, 110,000 people were detained because of their Japanese ancestry.[‡‡] Individualized determinations of national security threats were not made.[§§]

Herein lies the fundamental dilemma: on the one hand, American liberal values are essential to the American ethos; on the other hand, the combination of powerful strains of nativism with the residual effects of 9/11 and subsequent events, such as Northwest Flight 253, collectively and individually raises the profiling flag. While profiling—whether racially or ethnically—is at odds with the constitution and basic democratic principles, it represents an instinctual response to particular events. Whether it is effective is another matter. Therein lies the rub: while profiling does

---

[*] Berta Esperanza Hernández-Truyol, Natives, Newcomers and Nativism: A Human Rights Model for the Twenty-First Century, 23 Fordham Urb. L.J. 1075, 1089 (Summer 1996); Kenneth L. Karst, Belonging to America: Equal Citizenship and the Constitution 83 (New Haven, CT, 1989).

[†] Kenneth L. Karst, Belonging to America: Equal Citizenship and the Constitution 88 (New Haven, CT, 1989).

[‡] Dan Lacey, The Essential Immigrants 62 (1990).

[§] John B. Saul, Sound Ideas Obscured by Political Ideology, Seattle Times, October 30, 2005, at Sect. ROP Books.

[¶] Judge Andrew Napolitano as interviewed by Bill Steigerwald, Freedom, "the Default Position," Pittsburgh Tribune Review, October 8, 2005.

[**] Laura K. Donohue, Terrorist Speech and the Future of Free Expression, 27 Cardozo L. Rev. 233, 245 October 2005.

[††] David Cole and James X. Dempsey, Terrorism and the Constitution 150 (The New Press, 2002).

[‡‡] Id.

[§§] Id.

not reflect sound policy, it represents action that seeks to soothe an angry public, deeply concerned for its personal and collective safety.

However, that neither represents sound homeland security policy nor increases public safety. While, perhaps, a short-term political palliative, ethnic and racial profiling does not contribute to a safer America. How does that directly relate to homeland security in the context of immigration and narcoterrorism? While immigration represents the American dream, immigrants are perceived—both historically and presently—as the manifestation of uncomfortable dilemmas and problematic choices. This is no better articulated and manifested than in the public support of profiling in air security in the aftermath of Northwest Flight 253.[*]

NARCOTERRORISM

So, we circle back to an issue, previously discussed in this chapter, that cuts across many issues, some inherent to this book, others not. To that extent, when the book was initially conceived, narcoterrorism was not included in "issues to address." However, in the course of research (including interviews with subject matter experts), it became clear that a homeland security discussion analyzing immigration that does not include narcoterrorism is incomplete, at best. Complicating the discussion is a lack of a consistent and uniform definition of the term; that, however, must not serve as a barrier to discussing narcoterrorism. To that end, in an effort to forge ahead, narcoterrorism will be defined as the confluence between drugs and terrorism that simultaneously advances distinct and separate interests that merge to their mutual benefit.

Because individually and collectively the two are extraordinarily potent, their confluence (whether tactical or strategic) presents extraordinary homeland security threats. Drugs are a reality in America, various and strenuous efforts notwithstanding. While Nancy Reagan's "just say no" campaign received significant popular attention, and though the first Bush administration declared a "war on drugs" and appointed a drug czar, initially William Bennett and subsequently Bob Martinez, the reality is that drugs are inherent to the American culture. In the context of homeland security, the combination of drugs and terrorism brings to the fore two extraordinarily dangerous realities that pose a

---

[*] http://www.usatoday.com/news/nation/2010-01-12-poll-terrorism-obama_N.htm (last accessed April 10, 2010).

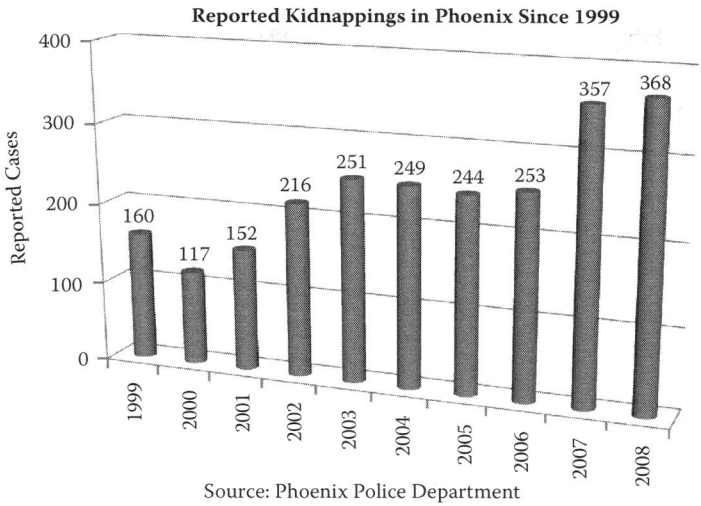

Reported Kidnappings in Phoenix Since 1999

Source: Phoenix Police Department

FIGURE 5.2 Kidnapping statistics for Phoenix, Arizona, since 1999.

present and a future threat alike. Simply stated: The enormous financial resources inherent to drug gangs and cartels directly contribute to unparalleled violence and fear in many quarters. While some observers may suggest that the combination of drugs, violence, and money is not a new reality, the concern is based on a significant increase in the violence and boldness of their actions.

In particular, the threat—previously alluded to—comes from Mexico. While experts interviewed for this book refused to brand Mexico a failed state, or even a failing state, they unequivocally articulated concern regarding the extraordinary violence of the Mexican drug gangs. What most assuredly deserves the attention of American decision makers is that the violence is not contained to Mexico; rather, it has become cross-border, with significant implications for American law enforcement officials. By example, the drug war-related kidnappings in Phoenix (357 kidnappings occurred in 2007 and 368 in 2008) may be but a precursor of the violence that may spill over into the United States (Figure 5.2).*

---

* http://www.cnn.com/2009/CRIME/05/19/phoenix.drug.kidnappings/ (last accessed December 20, 2010).

The violence—and the threat of increased violence, combined with enormous financial stakes—suggests that drug wars emanating from Mexico are distinct from wars between gangs with non-American roots. For example, although El Salvadorian gangs are notorious for significant levels of violence, their actions do not present a homeland security dilemma. This is distinguishable from the narcoterrorism threat posed by Mexican drug gangs as they potentially both threaten the stability of America's southern neighbor and impose their violence on American cities.

That combination, then, raises the banner of homeland security. To what end and to what degree is presently an "open question." Interviews with subject-matter experts made two things clear: the danger posed by both threats is palpable whereas the extent is unclear. The attention DHS Secretary Janet Napolitano directs to questions pertinent to Mexico and the U.S.-Mexico border are indicative of the concern the issues raise. This, in and of itself, is sufficient to raise the banner of homeland security.

6

The New Face of Terrorism
Domestic Threats and Civil Liberties

During the writing of this book I was invited to speaking engagements addressing a wide range of audiences, including corporate leaders, law enforcement officials, and academics and students in colleges, law schools, and public policy programs. The talks including frank exchanges during Q&A were exceptionally helpful to me in fine-tuning the issues addressed in this book. The audiences were disparate, reflecting distinct perspectives and concerns; accordingly, questions reflected those differences.

The one question I am repeatedly and consistently asked is whether domestic terrorism presents a viable threat to American homeland security. Sometimes the question is posed in an apologetic tone, sometimes with a sense of discomfort. Nevertheless, the question is posed so often by so many distinct audiences that it is clear to me that this is an issue very much on people's mind. Perhaps—with the exception of the economy—it is *the* issue on people's minds.

Certainly, particular events including the 42nd Street bombing, the Fort Hood bombing (see Figure 6.1), and the Islamic center/mosque controversy near Ground Zero highlight homeland security dangers and tensions. While this chapter focuses on actual attacks, the tone, tenor, and intensity of the national debate regarding the Islamic center is instructive, for it accentuates deeply held—and divisive—emotions regarding Islam and Muslims.

Deeply troubling and disturbing illustrations of underlying tensions predicated on religious-based hatred are Koran-burning ceremonies held

FIGURE 6.1 Soldiers attend a chemical light vigil held at the North Fort Hood training site on November 6, 2009, in remembrance of comrades and loved ones who were killed and wounded in the shooting tragedy at Fort Hood, Texas, the previous afternoon. (U.S. Army photo by Staff Sgt. Tony M. Lindback.)

by Christian churches[*] and attacks on mosques,[†] which has led to Muslims asking, as the *New York Times* defined the issue, "Will we ever belong?"[‡]

There is a sense, understandably, of discomfort with this issue, for it conveys, manifests, and projects images of McCarthyism, witch hunts, and careers destroyed. Images of Woody Allen's movie *The Front*[§] somehow seem appropriate and relevant. A valid concern is that in the current American political culture where discourse and debate are all but missing an issue sufficiently fraught with tension will become even more volatile, politically and publicly.[¶]

---

[*] http://wokv.com/localnews/2010/07/church-plans-quran-burning-on.html (last accessed December 20, 2010).

[†] http://www.foxnews.com/us/2010/08/09/texas-islamic-center-president-says-vandals-targeting-mosque-ground-zero/ (last accessed December 20, 2010). http://www.aljazeerah.info/News/2010/July/12%20n/Increasing%20Attacks%20on%20Mosques%20in%20the%20USA.htm (last accessed September 6, 2010).

[‡] http://www.nytimes.com/2010/09/06/us/06muslims.html (last accessed September 6, 2010).

[§] http://www.woodyallenmovies.com/movies2/the_front_zero_mostel.php (last accessed September 6, 2010).

[¶] http://stargtha.blogspot.com/2010/09/fwd-day-by-day-cartoon-by-chris-muir.html (last accessed September 6, 2010).

Nevertheless, the question is valid, legitimate, and on point. To engage in the discussion concerning the new face of terrorism requires discarding or at least suspending political correctness. Otherwise, the old military adage "we will be fighting yesterday's war" will be unfortunately applicable.

In the immediate aftermath of 9/11 the assumption oft-repeated in the media, scholarship, and government circles was that global jihadism posed the most direct threat to American national security. While understandable given 9/11, the discussion missed and continues to miss an additional, alternative paradigm. That additional paradigm—*domestic-based terrorism*—raises profoundly important criminal, constitutional, and moral questions, addressed below.

The possibility of domestic terrorism is disconcerting, for it reflects danger from within. Furthermore, it represents a perfect confluence—actually a clash—between security and liberty; because of the nature of the threat (internal), society must ask itself an uncomfortable question: What otherwise guaranteed rights are not to be extended to individuals living in our midst? Unlike international terrorism, domestic-based terrorism demands that we consider limiting protections otherwise extended to our fellow Americans, citizen and noncitizen alike.

It is important to distinguish between the different types of terrorists, such as:

- Terrorists from another country attack overseas (either U.S. interests overseas or other countries' populations, or both)
- Terrorists from another country come here to perpetrate attacks
- Terrorists (Islamic extremists) train overseas (or not) and perpetrate here
- Domestic terrorists: Christian extremist/neo-Nazi groups

Broadly speaking, I propose dividing homeland security into two distinct categories: the first is defined as Islamic extremism based on a combination of internal and external influences; the second is right-wing Christian fundamentalism predicated exclusively on domestic influences. The two, while distinct, share similarities: both are violent, uncompromising, on message, and mission dedicated, targeting innocent individuals living among their midst. A domestic terrorist has a much better chance of either knowing his victim or the victim's family than does an international terrorist.

Simply put: The 42nd street bomber (discussed below) statistically had a greater opportunity to have come in contact with his intended, random victims than did the 19 bombers responsible for 9/11.

93

This is precisely what makes domestic terrorism such an uncomfortable subject: the attacker may know his victim. While statistics demonstrate that in upwards of 80% of crimes committed the attacker knew the victim, the terrorism paradigm must be distinguished from the criminal law paradigm. The two are distinct; a terrorist's motivation is to advance a cause—religious, social, political, or economic—devoid of personal pecuniary gain or individual advancement. Conversely, the criminal law paradigm is largely predicated on an action that, ostensibly, benefits the actor, whether financially or personally, but is devoid of a cause, the essence of terrorism.

Simply stated: The traditional criminal does not have a higher purpose, whereas the terrorist is primarily motivated to advance a cause. The distinction is profound, affecting counterterrorism and traditional law enforcement policy, strategy, and means. Motivation is more palatable when its impact is felt thousands of miles away, affecting other people. While that sense of distance was shattered on 9/11, those responsible came from "elsewhere" and were, accordingly, not "of us." That is, while the terrorist attack occurred on U.S. soil, in the absolute epicenter of American financial and military power, its perpetrators were not our neighbors, work colleagues, and parents of our children's friends. While that did not make the attacks less violent or less horrific, there was a sense of remoteness perhaps—albeit unintentionally—best articulated in the oft-asked question: Why do *they* hate *us*?

Whether the question is appropriate or not has been discussed in the public and media; its relevance to our discussion is the they-us paradigm it articulates. Perhaps such an articulation made 9/11—in spite of its horror—more palatable, for we were not perpetrators but were only victims. As awful as 9/11 was, this significant distinction softened the blow. However—and this is the essence underlying the question I am repeatedly asked—the changing nature of terrorism suggests that "they" are "here," thereby hinting at a homeland security threat profoundly more distinct than previously envisioned and encountered.* That is, the terrorism paradigm has shifted from a threat perceived to be largely international or solely overseas in origin to increasingly domestic; in large thematic terms, 9/11 represented a new paradigm, whereas the increasing domestic-based threat manifests the newest paradigm.

From the perspective of homeland security, domestic-based terrorism imposes significant burdens and responsibilities that require a

---

* http://hoekstra.house.gov/UploadedFiles/NYPD_Report-Radicalization_in_the_West.pdf (last accessed September 5, 2010).

sophisticated—and expensive—combination of proactive and reactive measures alike.* As previously discussed, imposing (for lack of a better word) responsibility for 18 critical infrastructures on DHS is perhaps understandable from the perspective of the unfortunate current U.S. political environment and culture. However, it is unjustifiable and irresponsible from the perspective of those actually responsible for law enforcement. In simple terms: When asking senior law enforcement officials is it truly possible to address all 18 infrastructures, the honest response is a clear and resounding "no."

While the lack of prioritization reflects political reality, it is a disservice to the public and to homeland security officials alike; on the latter is imposed a mission best defined as impossible, whereas the former is seemingly ensured that security and protection are absolute, essentially hermetic.

Precisely because homeland security is neither absolute nor hermetic, prioritization as previously discussed is essential. That requires leaders—primarily officeholders and law enforcement officials—to engage in a sophisticated prioritization and resource allocation discussion. While perhaps painful, it is essential; otherwise, there truly will be no homeland security, much less a homeland security strategy.

The difficulty of this effort is compounded when the threat posed extends to deeply troubling events, including Hurricane Katrina and the BP oil spill in the Gulf of Mexico. Both examples, while resulting in enormous human tragedy and reflective of governmental incompetence and corporate greed, are profoundly distinct from the threat suggested in the chapter heading.

Perhaps because natural disasters are largely predicated on non-human influences, whereas domestic terrorism is exclusively based on human decision making, primarily (but not exclusively) influenced by religion, there are distinct differences in comfort level when addressing the two threats. The rational is undoubtedly easier to confront than the spiritual. However, to truly understand homeland security and to develop effective and efficient measures, the public and elected officials must assist law enforcement officials by understanding and addressing the threat posed by domestic religious extremism.†

---

* http://www.commentarymagazine.com/viewarticle.cfm/the-homegrown-terrorist-threat-15345?page=all (last accessed September 5, 2010).

† http://www.realinstitutoelcano.org/wps/wcm/connect/7b14d08040b690b988feda457bfe70e7/ARI171-2009_Vidino_Homegrown_Terrorist_Threat_US_Homeland.pdf?MOD=AJPERES&CACHEID=7b14d08040b690b988feda457bfe70e7 (last accessed September 5, 2010).

I have argued elsewhere that religious extremism poses the greatest danger both to democratic society* and to members of internal communities.† That is not to say that religious extremists in the United States threaten American national security, but it is to directly state that religious extremists living in the United States are a homeland security threat justifying concrete, proactive measures. There is, of course, a danger in identifying religious extremism as a viable threat and consequently proposing steps to counter the threat.

After all, both freedom of religion and freedom of speech are constitutionally guaranteed rights; the question, obviously, is: What if any limits can and should be imposed? Needless to say, the discussion regarding restricting free speech in the context of religious extremism is complex and controversial. That, however, does not suggest the conversation must not take place. The conversation is particularly complicated because limiting religious practice (by limiting free speech in the context of religious extremism) would, arguably, violate the Free Exercise Clause of the First Amendment. In the context of homeland security, the question is whether limiting otherwise constitutionally guaranteed rights is essential to more effectively protecting both external communities (the larger community) and internal communities (group-specific membership).

However, precisely because the threat posed by religious extremist-based domestic terrorism is neither vague nor amorphous, the proposed answer is yes. The question, obviously, is one of limits and proportion; disproportionate limitation of otherwise guaranteed rights would reflect arbitrary and capricious government action, be deemed void for vagueness and violating principles of overbroad regulation. There is, naturally, an important balancing discussion; after all, free speech is a sacrosanct principle.

In addressing the issue of free speech, the U.S. Supreme Court in *Brandenburg v. Ohio*‡ held that government cannot, under the First Amendment, punish the abstract advocacy of violence, thereby creating precedent with respect to speech protection. In reversing the conviction of a Ku Klux Klan leader who advocated violence after participating in a KKK rally, the Supreme Court held that government can limit free speech only if that speech promotes imminent harm, there is a high likelihood

---

* Amos N. Guiora, Freedom from Religion: *Rights and National Security* (Oxford: Oxford University Press, 2009).

† Amos N. Guiora, *Protecting the Unprotected: Religious Extremism and Child Endangerment*, 12 J.L. & Fam. Stud. 391 (2010).

‡ 395 U.S. 444 (1969).

that the speech will result in listeners participating in illegal action, and the speaker intended to cause such illegality.

The question is whether, in the face of contemporary threats, the *Brandenburg* test is too broad, imposing on the public an overtolerance* of free speech in the face of a direct threat. The question, obviously, is whether there is a direct threat justifying rearticulation of *Brandenburg*, or at least encouraging discussion whether the threat is viable, and what are the most effective means to counter it? Simply put, extreme religious speech presents a threat to individuals, internal communities, and society as a whole.

Precisely because of the danger presented by extremist religious speech, there is a compelling need to expand *when* religious speech may be restricted. As to the former, the discussion below highlights specific examples of homeland security threats predicated on religious extremism. One word of caveat: As Professor Fred Gedicks wisely commented, "Today's subversive speech may be tomorrow's conventional wisdom.... There is nothing more dangerous than giving the government power to punish speech based on who it decides is dangerous."[†]

Herein lays the critical question: How does society most effectively protect itself while preserving otherwise guaranteed constitutional rights? To fine-tune the issue: How does society protect itself against a threat (religious extremism) that goes to the core of the American ethos (freedom of religion/freedom of speech)? Extreme religious speech presents a threat to individuals, internal communities, and society as a whole. Precisely because of the danger presented by such extremist religious speech, there is a compelling need to expand **when** religious speech may be restricted.

The new face of terrorism can be concisely explained in the following manner: According to various reliable reports, throughout 2009 a number of second-generation Somali males in Minneapolis were radicalized in a mosque and subsequently went to Somalia to train to become suicide bombers.[‡] While uncertainty exists whether they were to conduct suicide bombings in the United States (either one or two conducted such an attack

---

[*] For an invaluable discussion of the tolerance-intolerance question, see Martha Minow, *Tolerance in an Age of Terror*, 16 S. Cal. Interdisc. L. J. 453 (2007).

[†] http://www.sltrib.com/sltrib/lifestyle/50044108-80/guiora-religious-particularly-religion.html.csp?page=4 (last accessed December 20, 2010).

[‡] http://www1.voanews.com/english/news/a-13-2009-03-27-voa43-68636162.html?CFTOKEN=57091416&jsessionid=663015a2bd228a5e3567537e66d163c201d3&CFID=256129143 (last accessed January 31, 2010).

in Somalia), it is clear that the imam and his radicalized congregants posed a direct threat to homeland security significantly facilitated by free speech protections, constitutionally guaranteed. While homeland security is not an absolute (policemen cannot stand at every corner), the question is: To what extent is society willing to protect itself in the face of a *known* threat?

In the context of proactive homeland security, it was incumbent upon law enforcement to take the necessary measures—once reliable information was received regarding the *essence* of this speech and the danger it *potentially* posed—to prevent the imam from continuing to radicalize his community. That, then, is the issue at hand: whether society with respect to homeland security is willing to prevent radicalization that can be directly attributed to cleric-based incitement.

Failing to do so facilitates incitement unencumbered by law enforcement. Advocates of strict application of the *Brandenburg* standard will argue that the decision *not* to limit the imam's speech reflects correct application of the holding. Conversely, if indeed home-grown homeland security threats are speech based, then arguably a requirement—going forward—is to minimize preexisting taboos, including immunity traditionally extended to religion, in particular religious extremism.[*]

The concern raised by the radicalization in Minneapolis cannot—and must not—be gainsaid in the name of political correctness, for the threat is palpable.[†] That said, the threat must not be exaggerated, as has been the traditional pattern in American history in response to direct or perceived threats. Those exaggerations have directly contributed to unwarranted impositions on political and civil rights. However, the danger posed by domestic religious extremism is compounded by a seeming unwillingness of senior policy makers to acknowledge the threat and the danger it poses.[‡]

The SUV in Times Square[§] discovered by alert t-shirt salesmen (military veterans) on May 1, 2010, was intended to be a major terrorist attack in the heart of New York City. Faisal Shahzad, the Pakistani-born, naturalized U.S. citizen, underwent training in Pakistan after becoming radicalized in his adopted country, thereby fitting the contemporary

---

[*] http://jurist.law.pitt.edu/forumy/2010/01/freedom-from-religion-learning-from.php (last accessed May 16, 2010).

[†] http://terrornewsbriefs.blogspot.com/2009/11/home-grown-islamist-terror-cells-in-usa.html (last accessed September 5, 2010).

[‡] http://www.ocregister.com/opinion/-248862--.html (last accessed May 16, 2010).

[§] http://www.nytimes.com/2010/05/02/nyregion/02timessquare.html (last accessed May 16, 2010).

radicalization paradigm manifested particularly in the United States[*] and the UK.[†] While Shahzad's mechanical competence may be an open question,[‡] thousands of innocent civilians— Americans and foreigners of all faiths—were deemed legitimate targets by Shahzad and his "senders."[§]

Given the potential, extraordinary harm the attack was intended to cause, Shahzad's arrest coming on the heels of the unsuccessful attack on Northwest Airlines Flight 253 directly led to intensive discussion regarding the appropriateness of extending Miranda[¶] protections to individuals suspected of involvement in terrorism.[**] The public debate regarding this possibility was addressed both in proposed congressional legislation[††] and in Supreme Court Justice Elena Kagan's confirmation hearings.[‡‡]

While Umar Farouk Abdulmutallab's attempt to blow up Northwest Flight 253 on Christmas Eve 2009 in Detroit, Michigan, was more akin to 9/11—a terrorist flown to the United States by senders to accomplish a single mission resulting in his death—than to Shahzad, there are important similarities. Both, after all, represent terrorism predicated on religious extremism with profound homeland security impact. The fundamental difference between Shahzad and Abdulmutallab is that the former was recruited overseas and had no—according to available intelligence information—prior connection to the United States, whereas the latter was a naturalized American citizen fulfilling the immigrants' American dream, referenced in the previous chapter.

Furthermore, in targeting innocent civilians, Shahzad and Abdulmutallab highlighted the soft underbelly of American homeland security: commercial airlines and the streets of America's cities. Abdulmutallab's attack was eminently preventable and reflects a systemic

---

[*] http://www.investigativeproject.org/documents/testimony/277.pdf (last accessed September 5, 2010).

[†] http://online.wsj.com/article/APb81bd0779d9e4dc39481c858ebfc3164.html (last accessed September 5, 2010).

[‡]

[§]

[¶] Miranda v. Arizona, 384 U.S. 436 (1966).

[**] See http://papers.ssrn.com/sol3/papers.cfm?abstract_id=1023101, http://papers.ssrn.com/sol3/papers.cfm?abstract_id=1622843, http://www.cambriapress.com/cambria-press.cfm?template=5&bid=319 (last accessed December 20, 2010).

[††]

[‡‡] http://lgraham.senate.gov/public/index.cfm?FuseAction=AboutSenatorGraham.Blog&ContentRecord_id=85b2e6d4-802a-23ad-48e2-b2596406c7de (last accessed September 5, 2010).

failing of unparalleled magnitude. While Shahzad's attack was similarly (thankfully) unsuccessful, it was extremely difficult to prevent from an operational counterterrorism perspective. In the case of Shahzad, it is important to note the motivation for his actions.

Shahzad's radicalization presents a significant, new challenge to law enforcement officials, for it was in part self-contained, reflecting self-radicalization, and in part based on the Internet, in particular the teachings of al-Qaeda cleric Anwar al-Awlaki.* President Obama has identified Awlaki—an American citizen—as a legitimate target for a drone attack.[†] Doubtlessly, Internet-driven incitement represents a twist on how faith leaders (and other inciters) can spread the message widely, quickly, and cheaply. In other words, dissemination is enormously facilitated by a means that law enforcement will be hard-pressed to negate.

Nevertheless, given the enormous power of both the Internet and religious extremism, mechanisms and countermeasures to minimize their influence are essential. Doubtlessly, the confluence is combustible, with extraordinary implications for homeland security and unparalleled complications for law enforcement officials. Essential to this effort is the understanding that religious extremist incitement clearly poses a significant threat to homeland security, and that limiting its impact is essential, even if the effort involves limiting a cleric's right to free speech.

Both Shahzad and Abdulmutallab, however, differ from U.S. Army Major Nidal Hasan, who deliberately targeted U.S. military personnel in Fort Hood, Texas, to protest deployment of U.S. troops to the Middle East.[‡] Hasan's actions manifest a slight twist from random acts of terrorism, in that they represent a category-specific act of terrorism targeting individuals who belonged to a particular, readily identifiable group, distinct from attacking a specific individual.[§] Whereas Abdulmutallab and Shahzad represent the mainstream of terrorist attacks—randomly choosing victims without concern regarding gender, ethnicity, or religion—for Major Hasan, killing American military personnel (his immediate peer group)

---

* http://newsbusters.org/blogs/alana-goodman/2010/09/01/youtube-jihad-american-terror-imam-radicalizing-muslim-youth-online (last accessed September 5, 2010).

[†] http://www.salon.com/news/opinion/glenn_greenwald/2010/04/07/assassinations (last accessed September 5, 2010).

[‡] http://www.politico.com/blogs/politicolive/1109/Lieberman_to_investigate_killings_at_Fort_Hood.html (last accessed May 16, 2010).

[§] An example would be political assassinations of a deliberately chosen target, such as a national leader.

was the means to articulate his deep opposition to American engagement in Afghanistan.

While many questions remain unanswered regarding law enforcement and military intelligence failure to identify Hasan as a legitimate threat prior to the attack, the process of his radicalization bears extraordinary resemblance to that of the Somalis[*] and some resemblance to that of Shahzad. In all three cases, the terrorists were radicalized by an imam, whether in a mosque or remotely (electronically). Particularly troubling, from the perspective of homeland security, is that Shahzad, Hasan, and the Somalis—unlike those responsible for 9/11 and Abdulmutallab—were not sent to the United States to commit a single act of terrorism, but rather, truly, lived in our midst.

The same seemingly holds true for three individuals placed in custody in the aftermath of Shahzad's arrest on the suspicion that they were responsible for providing him the funds necessary for the attack.[†] While the three, evidentially, were not U.S. citizens, and therefore held on civil immigration charges, it remains to be seen whether they, too, lived in our midst akin to Shahzad.

The train bombing in Madrid,[‡] the subway bombings in London,[§] the attack on the Glasgow airport,[¶] and the murder of Theo van Gogh[**] were all committed by nationals of each country, the same *modus operandi* adopted by Shahzad, Hasan, and the Somalis. The powerful combination of religious extremism premised on radicalization articulated in American houses of worship and the ability and willingness to travel to the respective family's traditional homeland (Pakistan and Somalia) for purposes of operational instruction directly contributed to a determined effort to kill innumerable innocent civilians.

The motivation for this new face of terrorism is largely religious extremism; in the case of the acts above, Islamic fundamentalism. That is not to suggest that religious extremists from other faiths do not commit acts of terrorism, quite the opposite.

---

[*] http://hsgac.senate.gov/public/index.cfm?FuseAction=Hearings.Hearing&Hearing_ID=70b4e9b6-d2af-4290-b9fd-7a466a0a86b6 (last accessed September 5, 2010).

[†] http://www.nytimes.com/2010/05/14/nyregion/14terror.html?nl=&emc=aua1 (last accessed May 17, 2010).

[‡] http://news.bbc.co.uk/onthisday/hi/dates/stories/march/11/newsid_4273000/4273817.stm (last accessed May 16, 2010).

[§] http://news.bbc.co.uk/2/hi/uk/4676861.stm (last accessed May 16, 2010).

[¶] http://news.bbc.co.uk/2/hi/uk_news/scotland/6257194.stm (last accessed May 16, 2010).

[**] http://dir.salon.com/news/feature/2004/11/24/vangogh/ (last visited May 16, 2010).

After all, abortion-performing doctors have been killed by Christian extremists in the United States,* Christian right-wing extremists have committed significant terrorist attacks† and planned (to various degrees) significant acts of violence, and a Jewish extremist, unequivocally incited by extreme right-wing rabbis, assassinated Israeli Prime Minister Rabin.‡ Targeting the physicians reflects category-specific terrorism (Hasan), whereas assassinating Rabin reflects person-specific terrorism distinguishable from random terrorism (Shahzad, Abdulmutallab, and McVeigh).

While extraordinary efforts, requiring unsurpassed resources, have been taken in order to prevent the next 9/11, the present, and arguably future, threat is fundamentally different. Rather than non-Americans coming to the United States to commit acts of terrorism, those committing the acts *are* American citizens. While distinguishing between Hasan and Shahzad, and Abdulmutallab is essential (American citizens/non-American), the emerging dilemma facing American decision makers is how to respond to both internal and external threats.

From the perspective of homeland security the dilemma is complicated, requiring sophisticated cost–benefit analysis based on an understanding of specific communities and religious motivation, the ability to translate coded religious speech, the willingness to consider imposing limits on otherwise guaranteed freedoms, including freedom of speech and religion, and the willingness to articulate what, today, presents an actual threat.§ The requirement, in the context of contemporary homeland security, is to both recognize and articulate the threat and to take decisive, proactive measures to prevent or at least minimize threats posed to homeland security by domestic, religious extremism.

To that end, developing a sophisticated intelligence network is essential. While difficult questions are inevitable monitoring and conducting surveillance of houses of worship reflects legitimate and legal law

---

* http://www.cnn.com/2009/CRIME/05/31/kansas.doctor.killed/index.html (last accessed May 16, 2010).

† http://www.time.com/time/2001/mcveigh/ (last accessed December 20, 2010).

‡ http://www.cnn.com/WORLD/9511/rabin/ (last accessed May 16, 2010).

§ In direct contrast to Attorney General Holder's comments in Congress, http://www.ocregister.com/opinion/-248862--.html (last accessed May 16, 2010), to Secretary of Homeland Security Napolitano's characterization of terrorism as "man-made terrorism," http://www.foxnews.com/story/0,2933,509597,00.html (last accessed May 16, 2010) and to President Obama's directive that particular words not be used in government documents, http://counterterrorismblog.org/2008/04/war_on_extremists.php (last accessed May 16, 2010).

enforcement. According to a recent RAND report,[*] U.S. home-grown jihadism increased threefold in 2009 but still remains marginal[†]; nevertheless, the challenge facing law enforcement is to identify the actors before they become operational.

To do so, in the context of constitutional limits imposed on law enforcement, requires determining the threat's primary source. By example: In a talk to state Homeland Security officials, I emphasized the essential requirement that they engage in intelligence gathering, including potential sources heretofore considered off-limits because of narrow-minded, short-term political considerations.

Because of the ethical concerns associated with conducting surveillance in houses of worship, appropriate probable cause standards must be determined relevant to churches, mosques, temples, and synagogues. Probable cause is based on the Fourth Amendment, which states that

> the right of the people to be secure in their persons, houses, papers, and effects, against unreasonable searches and seizures, shall not be violated, and no Warrants shall issue, but upon *probable cause*, supported by Oath or affirmation, and particularly describing the place to be searched and the persons or things to be seized.[‡]

While stereotypes such as "Muslims are dangerous" are clearly insufficient, the Supreme Court held in *Illinois v. Gates* that "probable cause requires only a probability or substantial chance of criminal activity, not an actual showing of such activity."[§] Thus, while the government cannot rely on stereotypes, once a probability of criminal activity can be shown law enforcement is justified in conducting surveillance. However, because of the danger of a chilling effect on the practice of religion, monitoring houses of worship must require a heightened probable cause.

That is, the traditional probable cause standard is insufficient for monitoring houses of worship because of the inevitable impact on the Free Exercise Clause. However, because of the danger posed by religious extremism—in particular incitement occurring *in* houses of worship—it is necessary to enable law enforcement's ability to monitor and conduct surveillance. While granting immunity to religion poses a clear danger to

---

[*] http://www.rand.org/pubs/occasional_papers/OP292/?ref=homepage&key=t_times_square (last accessed May 17, 2010).

[†] http://homelandsecuritynewswire.com/us-home-grown-jihadism-increased-three-fold-2009-remains-marginal (last accessed May 17, 2010).

[‡] U.S. Constitution Fourth Amendment, emphasis added.

[§] 462 U.S. 213, 245 (1983).

religion, the constitution cannot be used as a buttress to forbid the state from fulfilling its fundamental obligation. A heightened probable cause standard would resolve this tension.

ETHICAL CONCERNS IN CONDUCTING SURVEILLANCE

While it is clearly legal to conduct surveillance in houses of worship, concerns regarding stifling religious speech and religious worship require that significant attention be focused on *how* the surveillance is conducted. Deception is important to law enforcement in gathering information; the tension is between conducting surveillance openly at the risk of gathering less information and disguising law enforcement agents as worshipers who may be able to gather more information but do so in a deceptive manner.

Christopher Slobogin argues that law enforcement should not be able "to practice deceit in their official capacity, during interrogation or otherwise, unless (1) there is probable cause to believe the person to whom they are lying is a criminal; (2) the lying is necessary to obtain incriminating information; and (3) the lying does not have an illegitimately coercive effect."* If Slobogin's standards are followed, there is a profound question whether using deception in houses of worship is legitimate.

Slobogin admits that this first limitation "would curtail a significant amount of undercover work, pretextual seizures and searches, and lies aimed at witnesses and mere suspects. But it would also permit trickery during interrogation that follows an arrest, limited by the second and third requirement."† While interrogations are certainly important, undercover work is equally important and necessary for effective law enforcement. The second element, deception, is necessary to obtain incriminating information; without it, talk regarding criminal activity would almost never occur were readily identifiable law enforcement physically present.

Thus, the real question is whether it is ethical for law enforcement to deceive a group as a whole when only certain members are involved in the suspected criminal activity. As discussed below, deceptive law enforcement in monitoring a house of worship is ethical; however, in order to minimize the inevitable chilling effect, a balance between protecting society and protecting individual religious beliefs must be struck.

---

* Christopher Slobogin, *Lying and Confessing*, 39 Tex. Tech L. Rev. 1275, 1275 (2007).
† Id.

While resolving the tension between justified surveillance and the cost associated with such surveillance is a difficult issue, it is essential to adequately protecting the community. To that end, I recommend the following:

Enhanced cooperation between law enforcement and clergy would enable the former to warn the latter regarding suspected criminal conduct of individual congregants. Furthermore, where particular clergy are engaging in speech deemed capable of inciting, open channels of communication would facilitate law enforcement's ability to minimize a potential chilling effect by warning faith leaders of the potential criminal nature of their particular speech. This proactive discussion—warning faith leaders of speech that is possibly incitement—would at least lessen the need for future monitoring.

A heightened probable cause standard would enable monitoring of houses of worship while minimizing the chilling effect on people of faith. Determining whether previous speech justifies surveillance in accordance with a heightened probable cause standard would serve to narrow the instances of surveillance, ensuring that surveillance would occur only when and where truly required. This approach would significantly contribute to a more balanced and nuanced approach, for it would enhance law enforcement while protecting the freedom of religion. Rearticulated, a heightened probable cause standard would facilitate respect for the Free Exercise Clause while ensuring that government fulfills its primary obligation of protecting the public.

As previously mentioned, Christopher Slobogin argues it is deceptive for the FBI to conduct surveillance operations using undercover agents. A faith leader with whom I spoke indicated that were his church under surveillance, he would prefer FBI agents remain undercover to reduce the chilling effect. While arguably deceptive, it both leads to better information and minimizes Free Exercise violations. Were the FBI's surveillance efforts aimed at a particular parishioner, then the enhanced cooperation referenced above would be particularly important.

Finally, we must rearticulate the limits of speech as it relates to clergymen. How often does a clergyman need to incite before law enforcement moves in? What words justify monitoring? In *Brandenberg v. Ohio*, the court held that "the constitutional guarantees of free speech and free press do not permit a state to forbid or proscribe advocacy of the use of force or of law violation except where such advocacy is directed to inciting or producing imminent lawless action and is likely to incite or produce such action."* The court went on to say, "The mere abstract teaching of the moral propriety or even moral necessity for a resort to force and violence, is not the same as preparing a group for violent action and steeling it to such action."† The authority and power of an extremist religious cleric are, potentially, extraordinary. Therefore, when we examine the three prongs of the *Brandenburg* test—imminence, likelihood, and intent—the first two are almost certainly met in the case of an extremist religious authority determined to encourage his congregation to act. Sermons regularly addressing various dangers and evils will ultimately reach a "critical mass," and the listener's act will become imminent.

Respect for freedom of religion and understanding the danger of a chilling effect must serve as guides for *how* to conduct the surveillance. When monitoring is justified, it must be conducted both legally and morally. While immunity for religion ill-serves the state, respect for religion is the essence of civil, democratic society. To answer whether monitoring houses of worship is necessary, one must carefully examine the faith leaders' speech and conduct of parishioners. While turning a blind eye to religious extremism is an unaffordable luxury, chilling the practice of religion must be conducted with extreme care. Adoption of a heightened probable cause standard represents a legal and ethical solution to a most pressing issue.

INTELLIGENCE: FUNDAMENTALS

The intelligence community receives intelligence information from three primary sources: human intelligence, signal intelligence, and open-source information. To that end, intelligence information is defined, in part, as information concerning an enemy or possible enemy.‡ Determining if the information is "actionable" for operational purposes requires an analysis of whether it meets a four-part test: reliable, viable, valid, and corroborated.

---

* Brandenberg v. Ohio, 395 U.S. 444, 447 (1969).
† Id. at 448.
‡ Merriam Webster Dictionary, http://www.merriam-webster.com/dictionary/intelligence (last accessed March 22, 2010).

Test Prong	Definition/Use
Reliable	Past experiences show the source to be a dependable provider of correct information. Requires discerning whether the information is useful and accurate. Demands analysis by the case officer whether the source has a personal agenda/grudge with respect to the person identified/targeted.
Viable	Is it possible that an attack could occur in accordance with the source's information? That is, the information provided by the source indicates a terrorist attack that could take place within the realm of the possible and feasible.
Relevant	The information has bearing on upcoming events. Consider both the timeliness of the information and whether it is time sensitive, imposing immediate counterterrorism measures.
Corroborated	Another source (who meets the reliability test above) confirms the information in whole or part.

The danger posed by domestic terrorism is accentuated because the actor is difficult to easily categorize. Hasan was a major in the U.S. military promoted on a regular basis by his superiors; the second-generation Somalis faced the pressure of two distinct cultures with difficulties fitting in at school/socially; and Shahzad, the son of privilege (as was Abdulmutallab), was married with a family. The lack of uniformity with respect to class and life position presents law enforcement with an extraordinary challenge; the threat is not posed by individuals residing in one social stratum.*

Precisely because U.S. law enforcement—with specific and tragic exceptions†—does not engage in a "round up the usual suspects" approach, *proactively* identifying the next domestic terrorist requires understanding what is the unifying theme. Simply put: What are the constants that tie these individuals together who come from different parts of the country and from distinct life stations?

---

* By comparison, those responsible for the attacks in the United Kingdom came from upper-middle-class backgrounds in different cities, whereas those responsible for the riots in France (2005) shared common geographic and social backgrounds; http://riots-france.ssrc.org/ (accessed May 25, 2010).

† See *Korematsu v. United States*, 323 U.S. 214 (1944); The Brig Amy Warwick (The Prize Cases), 67 U.S. (2 Black) 635 (1863); Attorney General A. Mitchell Palmer on Charges Made against Department of Justice by Louis F. Post and Others, Hearings before the Committee on Rules, House of Representatives 27 (1920) Palmer Raids; Alien and Sedition Acts: Naturalization Act, ch. 54, 1 Stat. 566 (1798); Alien Friends Act, ch. 58, 1 Stat. 570 (1798); Alien Enemies Act, ch. 66, 1 Stat. 577 (1798); Sedition Act, ch. 74, 1 Stat. 596 (1798).

The motivation in these cases is, broadly speaking, religious extremism fueled by a rigid understanding and interpretation of religious text reflective of a "call to action" that brooks no dissent. Whether the actor is radicalized in the mosque (as was the case with the Somalis) or electronically (as was the case with Hasan), the result is the same: terrorism in the name of religious extremism. Regarding Hasan, the reality, from a homeland security perspective, is that a senior officer in the U.S. military was sufficiently radicalized to act in the name of religious extremism.

To that end, I propose the following eight steps:

1. **Local-state-federal cooperation:** Cooperation and coordination between governmental agencies within, and among, the local, state, and federal governments are essential to maintaining an effective homeland security strategy.
2. **Private-public sector cooperation:** The private sector, which plays a vital role in the homeland security plan, must coordinate with the public sector to move toward a viable private-public sector initiative.
3. **Limits of interrogation:** While interrogation practices represent a crucial meeting ground between human rights and counterterrorism measures, the limits placed on interrogators are perhaps the most difficult to define, for they determine how far a civil society is willing to go in fighting the exigencies that terror presents.
4. **Balancing personal privacy with national security considerations/limits of civil and political rights:** Finding a balance between national security and the rights of individuals is the most significant issue faced by liberal democratic nations in developing counterterrorism strategy. Without a balance between these two tensions, democratic societies lose the very ethos for which they fight.* Indeed, it is imperative for democracies to avoid infringing on political freedoms and civil liberties. Yet, the ultimate responsibility of government is to protect its citizenry. The struggle to balance competing interests is the fundamental dilemma confronting democracies today.

---

* Benjamin Franklin, Pennsylvania Assembly: Reply to the Governor, Nov. 11, 1755, in The Papers of Benjamin Franklin, ed. Leonard W. Labaree, vol. 6 (1963), 242.

5. **Interacting with the public** (cultural, religious, and ethnic considerations and sensibilities): In order to prevent loss of innocent life, violations of human rights, and waste of public funds, we must teach/train homeland security officials cultural norms and sensitivities and language and communication skills.

6. **Arrest and remand of individuals suspected of involvement in terrorism:** Effective counterterrorism strategy must be based on sophisticated risk assessment; to that end, an equal risk approach with respect to individuals suspected of involvement in terrorism is operationally unfeasible and ineffective, which requires law enforcement officials to consistently seek more sophisticated and accurate intelligence information.

7. **Religious extremism/freedom of religion:** Incitement by extremist Islamic clerics in recent years is a growing area of concern for members of the intelligence community. To that end, relevant questions include:
   When should First Amendment limitations apply to faith leaders? How should U.S. intelligence reporting properly address/caveat religious rhetoric vs. criminal intent?

8. **Freedom of speech:** Homeland security debates have addressed what civil liberties should be honored, including torture, domestic surveillance, and unlawful detentions; addressing the limits of freedom of speech is essential to local and national law enforcement.

7
Terrorism Financing

INTRODUCTION

Money is the true lifeblood of terrorism. When discussing homeland security and how we use threat analysis to protect the nation's life and property, significant resources should be directed toward tracking dollars, rather than bombs. By the time the explosives have been purchased and assembled, the game has sufficiently changed to a dangerous defensive posture. However, a proper threat analysis will move the point of examination earlier, where officials will not solely be searching for the bombs, but rather will be looking at, and trying to prevent, the earlier transactions that funded and purchased the explosives—effectively stymieing the terrorists' capabilities to finance their operations. If the United States is successful at slowing, or stopping, the flow of terrorism financing, the number of attacks will fall precipitously; without financing, the triggermen will have no, or at least diminished, resources with which to work.

There may be hundreds of men and women willing to carry a bomb, but operationally, eliminating one of them merely makes room for another. However, only a small number of people act as financiers of such attacks, and they have a much greater overall impact on terrorism than mere bomb carriers. As such, the bull's-eye of counterterrorism must be expanded to larger concentric circles that include not only the fighters, but also those providing material support. This discussion does not argue for the killing of such financiers, but rather for an acknowledgment that these individuals must be pursued with the same intensity as the bomb carriers themselves. While the use of military and law enforcement in counterterrorism operations achieves "on the ground" objectives of rooting out terrorists, legislators must take greater steps to permanently close the loopholes easily used by unscrupulous investors.

In order to fully understand terrorism, it is a pivotal presupposition that finances are the engine of the terrorist train. However, notwithstanding the financier's extraordinary power to enact terrorism, in taking proactive steps against terror financing, governments must recognize and balance equally imperative considerations. The need for a thorough understanding of what qualifies as homeland security policy requires relevant policy makers operate under a precisely defined purpose, so as to properly balance policy decisions against an individual's freedoms. Only when the government has a well-defined understanding of where it can and cannot act is it able to both protect the nation's safety and simultaneously ensure its citizens the rights protected by the Constitution.

While the PATRIOT Act, specifically its amendments to the Bank Secrecy Act, has provided the government with many effective tools with which to combat terrorism financing, it is still insufficient for two reasons. First, while procedures and rules are important and helpful, the issue turns on application—an area where the government is still lacking, particularly in the context of this discussion where there is a complete void of consistent definitions of what is and what is not within the umbrella of homeland security as it is applied to terrorism financing. Second, the new rules and regulations insufficiently quell the informal value transfer systems.

Lest anyone think that the U.S. government has successfully addressed terrorism financing following 9/11, the attempted attack in Times Square should stand as a reminder of just how wrong such a conclusion is. For example, reports that followed the May, 2010 attack indicated that U.S. authorities were looking for a money courier who may have helped the alleged attacker purchase the SUV that was used in the attempted attack. Specifically, the authorities were looking into allegations that an individual funneled several $100 bills to the attacker in order to purchase the SUV without being connected to the financiers in question. Stemming from this report, various news sites have reported efforts made to combat terrorism financing. One such CNN report discusses the dispatch of U.S. Treasury officials to Kabul to attempt to track money laundering schemes through large, informal banking networks using annual income generated from the Taliban's involvement in the illicit drug trade.

In addition terrorists have employed many other methods of funding their activities. Including investments, banking systems, and particularly the informal value transfer system of hawalas (which uses many techniques similar to those used in money laundering).

WHAT IS TERRORISM FINANCING?

Terrorism financing mirrors and implements many characteristics found in money laundering. Money laundering is the "process by which one conceals the existence, illegal source, or illegal application of income, and then disguises that income to make it appear legitimate." This "cleansing" of money's illegitimate source has long been a mainstay of criminal activity in the United States, as it facilitates hiding criminally derived proceeds.

The traditional method of laundering money requires the successful completion of three steps: (1) the launderer must place the "dirty" money into a legitimate enterprise; (2) these monies are layered through multiple and separate transactions so as to obscure the origins of the money; and (3) the now "clean" money is brought into the legitimate financial community through bank notes, loans, or other market-based instrumentalities. This discussion, however, becomes all the more important—and dangerous—when taken beyond the domestic criminal context into the international world of financing terrorism.

The discussion of money laundering and the governmental reaction thereto is important for identifying both the successes and failures of curbing terror financing. While there are differences between traditional money laundering and terrorism financing (most notably, the fact that much of the terrorism financing money does not have an illicit or illegitimate origin), the similarities in methods and mechanisms of moving money allow for terrorism financing lessons to be drawn from the U.S. government's long-standing fight against money laundering. This preliminary discussion is also valuable because prior to 9/11, anti-money laundering legislation represented the sole weapon to stop terrorism financing, a set of laws that the attacks of 9/11 showed to be inadequate.

Terrorism financing is frequently implemented through the transnational transferring of money and property. Specifically, the use of informal value transfer systems (IVTSs) is commonly referred to as underground banking because, although operating akin to a banking system, the IVTS does so without participating in the formal requirements of institutional banking. However, calling such networks underground banks does not accurately portray the operation of an IVTS. The IVTS is primarily a system for the transfer of money and assets rather than an actual provider of full banking services. These networks are commonly used in terror financing for their ability to move funds around the world without the actual movement of a single traceable dollar.

Adding to the complexity of finding improper uses of IVTS networks, these networks are oftentimes operated openly and legally, as there is no illegality involved in solely transferring value and the systems that often provide a valuable and necessary legitimate service to many people. Making these networks even harder to locate and monitor, most IVTS agents operate numerous legitimate business ventures. Creating another layer in this discussion of how the government ought to respond to abuses of the IVTS system without infringing on the protected freedom of religious exercise, in many nations such networks are the sole means of value transfer often used out of religious duty.

Specifically, an IVTS operates not by exchanging money, which would be traceable, but rather through the exchanging of debts, where the only tracking method is a balance sheet. In transacting these debt transfers, an IVTS agent will often use untraceable actions, like false pricing on imports or exports, in-kind payments, trade diversion schemes, or the use of prepaid phone cards.

Prior to 9/11, money laundering laws were the spear's tip for fighting terrorism financing. The pervasiveness of money laundering in the United States during the 1980s brought about the Money Laundering Control Act of 1986. The act was intended to establish liability for an individual who conducts a financial transaction with knowledge that the funds' origins are either illegal or illicit. This act represented the government's initial action aimed at the money launderer specifically, as prior governmental enactments focused on the movement of illicit monies by financial institutions and often overlooked the individual altogether.

The 1986 act also served to specifically define what acts would constitute money laundering. Specifically, the legislation includes previously used definitions of income from legislative acts that responded to organized crime, including prostitution, gambling, drug trafficking, and violations of the Racketeer Influenced and Corrupt Organizations Act. The 1986 act further broadens the definition of money laundering to include proceeds from copyright infringement, espionage, trading with the enemy, and violations of the Internal Revenue Code.

For the discussion of terror financing, the 1986 act offers an initial lesson showing that definitions must be broad so as to impact all actors.

Further, the 1986 act provides another important example for this discussion, as it not only bans the specific criminal act of laundering money, but also has much greater impact by (1) making illegal any use of such funds (2) in perpetuity, without a statute of limitations.

After the 1986 act, but before 9/11, attacks morphed the money laundering issue into a terror financing issue; enforcement mechanisms against money laundering were found in 18 USC §§1956–57. However, despite the success of this legislation against money laundering, these sections raise concerns for the fight against terrorism financing in a post-9/11 world.

Section 1956(a)(1) was enacted to focus on the transactional aspect of money laundering, where the statute only applies if the transaction specifically handles monies received from illegal ventures. Thus, although this statute may be effective in the campaign against typical crime-related money laundering—as such money is usually derived from the sale of drugs, prostitution, or gambling—it raises concerns for fighting terror financing. Specifically, this provision is inadequate in stopping terror financing, as it often involves an individual independently giving his or her personal funds (which are fully legal monies) to another person who may eventually fund terrorism. In this type of terror financing transaction, where legal monies move between parties, §1956(a)(1) would never be triggered because the money being moved is legal, or clean, at the time of the transaction.

The 1986 act then specifically addresses the transfer of funds under §1956(a)(2). Under 18 USC §1956(a)(2), the act of transferring illicit money in or out of the United States is illegal. This section of the statute is a more powerful weapon than §1956(a)(1), as §1956(a)(2) does not require the showing that the monies be direct proceeds of an illegal action. However, despite this more powerful weapon against money laundering, §1956(a)(2) still illuminates lessons for future terror financing legislation because (1) §1956(a)(2) still requires a transfer, and there are questions of whether a transfer is found in value transfer systems where no actual money moves, and (2) §1956(a)(2) has a provision mandating that the money cross the border of the United States. The implication here is that §1956(a)(2) does not become active if the funds move only within the United States, or if they move only outside of the United States.

Requiring money to cross the American border before jurisdiction is effectuated raises concerns in fighting terror financing, as financiers can effectively avoid §1956 by moving legal money domestically or internationally. As such, the recommendations at the conclusion of this article mandate that the United States gains jurisdiction over any money transfer aimed at the furtherance of terrorism.

The 1986 act provided that the U.S. government, in prosecuting an individual under the 1986 Money Laundering Act, must satisfy four elements of the crime: (1) knowledge, (2) the existence of proceeds derived from a specified unlawful activity, (3) the existence of a financial transaction, and (4) intent to launder money. Although the Money Laundering Act requires some form of knowledge, the specific type of knowledge varies by specific offense. In general, the government must show knowledge that there was some sort of unlawful underlying transaction that led to the money at issue, while some circumstances require the more specific knowledge of the exact unlawful activity. Of particular importance in establishing new rules for terror financing, the question of willful blindness was left unanswered.

However, the knowledge requirement most important for this discussion is whether willful blindness can stand for knowledge. Both §§1956 and 1957 require "actual knowledge," a more stringent standard than a negligence theory of should have known or reckless disregard. In order to reconcile this issue, courts have taken proactive steps to erode this hardline rule of requiring actual knowledge by a finding that knowledge may be satisfied through willful blindness.

Willful blindness is vitally important to terror financing, as an individual may send his clean money to a person whom he does not know for sure will use the money for terrorism, yet knows that the recipient has on multiple previous occasions funded terrorism. Thus, as terror financing is a more evasive system than strict money laundering, the open-ended definitions used in anti-money laundering legislation cannot double as definitions in terrorism financing legislation. In promoting new and effective methods of curbing terror financing, willful blindness must be statutorily held as tantamount to actual knowledge.

The second element of a money laundering offense requires showing an action that involves the "proceeds of specified unlawful activity." The first part of this requirement is the definition of the term *proceeds*. How tangential can income be and still be considered proceeds? Or, more importantly, how far back must the government trace money to find the money at issue to be proceeds? Unfortunately, under §1956, the term *proceeds* is not sufficiently defined, and under §1957, the statute merely uses "criminally derived property" to stand for *proceeds*. When the transaction changes from a money laundering campaign involving prostitution proceeds into the world of terror financing, where the financial transaction supports terrorism, a statute cannot have an element of the crime that is left undefined and open for legal argument and maneuvering.

In addition, another important aspect of the money laundering rules, which highlights better ways to handle terrorism financing, is the unanswered question of how far back the government must trace money to find proceeds. Under §1956, the government does not necessarily have to trace the dollar to a particular offense. Rather, the government is only required to establish that the defendant participated in actions that are typical of criminal activity, and that there was no other legitimate source of the funds. However, the legal system has been reluctant to permit such circumstantial showings to stand wholly on their own, preferring such showings only to allow a jury to make an inferential finding that there could not have been a legal source of the funds.

Thus, the fact that the government need not actually prove the predicate offense makes this statute an advantageous weapon for government prosecutions. Section 1956 further defers to the government's case, as the government is not required to trace the funds where an individual is shown to have intertwined the illicit funds with legal income. Thus, as noted above, in order to sustain its burden, the government need only show that a portion of the funds in question was more likely than not involved in illegal activity.

Beyond the deference to the government regarding standards of proof, the topic of commingled funds is also highly deferential to the government in enforcement, as a conviction involving commingling of funds will result in the forfeiture of all funds (no matter which parts are legitimate). For homeland security, definitions are extremely important, and it is important that the laws used to combat terror financing specifically define the government's duty in tracing money.

Further, this section requires those proceeds to be of a "specified unlawful activity." The 1986 act itself offers an expansive list of specific crimes that will satisfy this requirement. While the enunciation of specific underlying crimes may be sufficient against money laundering, such restrictive language was inadequate in stopping the terrorist attacks of 9/11. Specifically, money laundering focuses on the cleansing of illegal monies, and thus creating a list of underlying offenses will permit effective prosecution. However, in terror financing there is often no illegal underlying offense, but rather a future intent to use the funds illegally. Thus, a requirement that the money have a specific origin is unnecessarily restrictive in fighting terror financing.

The third requirement for finding a money laundering offense is the existence of a financial transaction. Contrary to common understanding, a financial transaction is not limited to merely banking or investment house transactions. Rather, the statute's use of the term *financial institution* creates liability for any exchange of money between two parties, so long as the transaction in some way impacts interstate commerce and meets one of the four intent requirements found in §1956(a)(3). Specifically, to be a violation under §1956, the activities must impact interstate commerce or involve a "financial institution which is engaged in, or the activities of which affect, interstate or foreign commerce in any way or degree." This requirement is not as important for the actual classification of an action as constituting money laundering per se; rather, the standard exists so as to effectuate federal jurisdiction. However, such a standard is easy to meet, as courts are often lenient in finding an impact on interstate commerce, permitting such a finding with only minimal effects. Thus, in seeking principles from money laundering legislation to apply to terror financing legislation, this requirement does not raise substantial issues for litigating terror financing.

The fourth legislative requirement for money laundering is intent. The methods for showing intent under §1956 are finding the:

(A) (i) intent to promote a specified unlawful activity, or (ii) intent to engage in a violation of the Internal Revenue Code,

(B) knowing that the transaction is designed in whole or in part (i) to conceal or disguise the nature, the location, the source, the owner-ship, or the control of the proceeds of specified unlawful activity, or (ii) to avoid a transaction reporting requirement under State or Federal law.

Alternatively, a §1957 prosecution only requires a showing of knowl-edge that the financial transaction is occurring, without specifically requiring the intent to launder money, making §1957 a more advantageous weapon for prosecution.

Showing intent, much like showing knowledge, requires a fact-spe-cific analysis. The intent requirements under the 1986 act demand that a defendant acts knowing that the transaction, or movement of property, is designed to hide information about the proceeds of the specified crimi-nal activity. As referenced earlier, the intent for terror financing must be broad enough to include any financial act intended to further terrorism. But, throughout this discussion, the question remains: How does this area of anti-money laundering tie to terror financing and the modern-day war on terrorism?

TERROR FINANCING—9/11 AND BEYOND

The legal and political impact of the 9/11 attacks not only altered national security and international law, but also impacted relevant legislation. The specific alteration at issue in this discussion is the legal and policy implications on financial transactions after 9/11. Historically, the use of IVTS networks has been tied to kidnapping, tax evasion, corruption, and weapons smuggling. More important to this immediate discussion, however, is that IVTS and terrorism financing have been used both for the 9/11 attacks and more recent events.

Before engaging in the discussion of terror financing specifically, it must be highlighted that terrorism is fully dependent on money and financiers. Efforts to change, or "win," the hearts and minds of terrorists are important, but pose limited likelihood of success on their own. Rather, it is a more powerful counterterrorism weapon to cut off the lifeblood of these individuals, making their mindsets a moot point.

Despite many similarities, terror financing presents a wholly different discussion from money laundering, and as such, the traditional money laundering legislation is insufficient against terror financing. Money laundering, as the above discussion suggests, is the cleaning and concealment of dirty or illicit money. A government program that searches for illegal activities will likely find money laundering. However, the financial acts in terror financing do not necessarily involve illicit funds. In money laundering the criminality begins with the illicit earning of funds, followed by the subsequent illegal act of money laundering. In terror financing, however, the actual illegality often occurs only after the actual transfer, when the money is ultimately used for funding terrorism. Thus, the mere application of the existing money laundering rules is insufficient.

In short, the problematic nature of IVTS networks is that it is practically impossible to track the funds due to the fact that most dollars passing through an IVTS are legitimate and clean. In the IVTS networks, clean money is sent through a system populated by mostly clean money, and the funds reach their illegal purpose when used for terrorism.

The United States has promoted many different agenda points aimed at curbing money laundering internationally in an effort to make it less enticing for domestic individuals to use the international markets to launder money and finance illicit actions. One such effort was the creation of the Financial Action Task Force (FATF), the investigative body of the Organization for Economic Cooperation and Development. FATF promotes U.S. interests by requiring nations to institute their own domestic legislation in compliance with established regulations. Beyond tacitly requiring countries to create their own domestic laws aimed at money laundering, FATF puts forth special recommendations that specifically delineate legislative actions nations are urged to follow.

Special Recommendation VI addresses IVTS networks; specifically, it states that countries must "license[] or register[]" all informal value transfer businesses and subject them to the same FATF requirements as banks and financial houses. This recommendation has significant implications, as the failure of a nation to comply can result in the G-8 nations adding the noncompliant nation to a blacklist of noncooperating countries and territories until that nation agrees to comply with such recommendations.

The United States has stepped further into the international realm by specifically identifying a "hawala triangle" between Dubai, Pakistan, and India, as they are the areas most heavily invested in hawalas. The United States became particularly interested in this area given that Mohamed Atta and Marwan Al Shehhi, two of the 9/11 hijackers, received more than $120,000 from Dubai in 2000.

Responding to the U.S. scrutiny of the hawala triangle, the United Arab Emirates took measures in the international community by participating in a 2002 financial conference with more than 300 international delegates. At this conference the Abu Dhabi Declaration on Hawala was adopted. The goal of this declaration was to articulate and recognize the positive aspects of hawalas as they "provide[] a fast and cost-effective method for worldwide remittance of money," while also calling for their effective, but not overly restrictive, regulation. More important, though, was the declaration that the "international community should remain seized with the issue and should continue to work individually and collectively to regulate the Hawala system for legitimate commerce and to prevent its exploitation or misuse."

In order to further ascertain the effective and ineffective ways of targeting terror financing, the discussion of laws on the books before the 9/11 attacks must move beyond the money laundering-specific legislation to a discussion of other efforts aimed at regulating the illegal use of the IVTS. Specifically, the Bank Secrecy Act of 1970 (BSA) was the initial legislation requiring record keeping and reporting requirements for banks. Then, Congress specifically addressed IVTS networks through the aforementioned Money Laundering Control Act of 1986, which includes the more often recognized money laundering rules. In attempting to coalesce these standards with counterterrorism efforts, the U.S. Congress legislated the United and Strengthening America by Providing Appropriate Tools Required to Intercept and Obstruct Terrorism Act of 2001 (PATRIOT Act). The PATRIOT Act is specifically applicable to this discussion, as it makes the failure to comply with the BSA's reporting requirement a criminal, rather than merely a civil, offense. Additionally, it amends the BSA to specifically address IVTS networks.

Beyond these legislative responses, the U.S. Department of the Treasury established an enforcement division called the Financial Crimes Enforcement Network (FinCEN), which works with various U.S. law enforcement agencies in an effort to ensure compliance with the above legislation. While FinCEN focuses on domestic enforcement, the international realm is covered by the Office of Foreign Assets Control, which focuses on disrupting and freezing illicit funds internationally. The enforcement "teeth" beyond these legislative regimes and entities exist in the Department of the Treasury's interagency enforcement group, Operation Green Quest, which is a "multiagency task force led by the U.S. Customs Service that also includes the Internal Revenue Service, the Secret Service, Treasury's Office of Foreign Asset Control, and FinCEN." This multiagency task force operates to "augment existing counter-terrorist efforts by bringing the full scope of the government's financial expertise to bear against systems, individuals, and organizations that serve as sources of terrorist funding."

Beyond the aforementioned governmental responses, there have also been significant statutory efforts to curb terror financing. However, although the federal government has pursued substantial efforts regarding terror financing, it simply needs to go further. The most direct and effective measures on point are the material support statutes. Despite the value they serve in prohibiting material support, they are insufficient because of two shortcomings.

First, Sections 2339A and 2339B both make it unlawful to knowingly or intentionally provide resources to terrorists or terror organizations. However, the knowledge requirement leaves an impermissible door open whereby an individual insulates himself from culpability by acting in a "willfully ignorant" manner. Thus, to cure this deficiency, the statute must apply to providing any material support.

Second, these statutes necessitate the transfer specifically to a terror organization. However, such an element is extraordinarily difficult to show. The statute must be expanded to outlaw material support while clearly delineating specifically what is required for showing the connection of support.

Hawalas are an often discussed and criticized source of terror financing. The IVTS system is popular because of its low fees and lack of formalities. However, the lack of formalities raises the danger that they will be used for illicit purposes, since there is oftentimes no paper trail of a transaction, the money does not cross the American border, and the money is legal, or clean, at the time of the transfer. This is dangerous considering, for instance, that the primary focus of legislative efforts to curb money laundering and terror financing before 9/11 was on (1) the illegal nature of the money in question and (2) the need for such money to cross the border. Neither of these two triggers is activated in the hawala example below. As a primary method of IVTS, the hawala is used around the world to transfer money or assets without either a paper trail or the high fees charged by banks. In order to understand how an IVTS system is used to finance terrorism, it is imperative to see how such a transaction occurs and to see the complete lack of formalities or paper trails. The following illustration will describe how such value transfer systems work.

An American citizen (AC) wants to send $1,000 to his friend (F) in Turkey. AC contacts a hawaladar (H1) in the United States to effectuate this transfer. H1 consents to make this value transfer from AC to F for a 2% fee, an amount less than charged at banks or wire transfer businesses. AC then pays H1 $1,000 and H1 gives AC a password. After this, AC contacts F to give him the password and tells F whom to contact in Turkey to receive the money. At the same time, H1 contacts his business partner, a hawaladar in Turkey (H2). H1 informs H2 of the transaction and H2 gives the same password H1 gave AC. When F meets H2 and gives H2 the password, F receives the local equivalent of $1,000 minus the 2% commission. At no point did an actual dollar move between countries in this transaction.

However, this is only half of the transaction. While AC and F have completed their transaction, H1 received $1,000 and H2 paid the local equivalent of $1,000 minus the commission fee. Thus, there is a debt of roughly $1,000 between hawaladars. One way to repay such debts is through reverse transactions where a person in Turkey wishes to send $1,000 to a person in the United States and opts to use H2 and H1 for such a transaction. However, another commonly used method between hawaladars is through legitimate business. When the hawaladars are involved in importing and exporting, for example, the $1,000 debt can be repaid by adjusting an invoice to overstate the value of the legitimate goods by $1,000. Or, alternatively, H2 may owe $1,000 to another hawaladar in America, and in order to satisfy the debt incurred in the transaction described above, H1 may pay H2's debt to the third party.

This brief picture of the hawala system shows how the achievement of a simple goal requires a complicated set of transactions. This picture also highlights how hawalas work without any physical transfer of money between primary parties and is based fully on trust and obligations rather than paperwork, making regulation and tracking very difficult. The nature of the hawala system, and its potential for abuse, requires law enforcement to address regulation of the system. Although it is nearly impossible to gauge the size of hawalas worldwide, it is estimated that hawalas involve billions of dollars traveling around the world through these informal, unregistered networks.

Under the aforementioned legislative framework of money laundering statutes and the PATRIOT Act, the U.S. government aims to regulate illegal money transfers through IVTS. Specifically, the PATRIOT Act acts as a way to patch many of the previously discussed holes through which terror financiers have slipped. Under the PATRIOT Act, all individuals or entities that transfer monies, no matter how formal or informal, must comply with the aforementioned money laundering regulations. For instance, 31 CFR §103.20 requires money service businesses to comply with the suspicious activity reports requirements imposed on other financial institutions. Of the most importance to this discussion, the PATRIOT Act redefines "money transmitting business" to include any person "who engages as a business in the transmission of funds, including any person who engages as a business through an informal money transfer system or any network of people who engage as a business in facilitating the transfer of money domestically or internationally outside of the conventional financial institutions system." The importance of this definitional change lies in that the strict reporting rules previously applicable only to banks now apply to these informal transfer entities.

Under the BSA, the term of art used for these nonbank transfer entities is money transmitters or money service businesses (MSBs). The PATRIOT Act applies the financial rules to IVTS networks by making it illegal to run an MSB (1) without the applicable state license, (2) outside of FinCEN's requirements for registration, or (3) with knowledge that the money had an illegal origin. The origins of the PATRIOT Act defined "unlicensed money transmitting business" as any "money transmitting business which affects interstate or foreign commerce in any manner or degree and (A) is operated without an appropriate money transmitting license ... (B) fails to comply with the money transmitting business registration requirements ... or (C) otherwise involves the transportation or transmission of funds that are ... intended to be used to promote or support unlawful activity."

The PATRIOT Act strengthened the financial enforcement system by requiring every MSB to register like a corporation, providing contact names and numbers, which must include an owner's name, his contact address, the MSB's financial account numbers, and a number given it by the federal government. This is important to the discussion at hand because, as noted earlier, the lack of formalities in IVTS transfers is the primary reason for the ease with which they are abused. Imposing reporting requirements similar to those applied to banks was an attempt to eliminate loopholes.

The first PATRIOT Act case study of government action against an MSB involved Mohamed Hussein. Hussein ran an MSB without having complied with the licensing requirements and was sentenced to 18 months in jail. This case highlights an important precedent because Hussein argued a "lack of knowledge" defense, which the court held to not be viable. This holding is important to the fight against terrorism financing, for denying an argument based on lack of knowledge will preclude an investor from being willfully blind about the destination of his funds.

In a second case, al-Barakaat was subject to a PATRIOT Act investigation. Al-Barakaat was a financial entity based in Somalia operating throughout Europe and North America. Further, al-Barakaat was the largest employer in all of Somalia. In addition, al-Barakaat was also believed to have funneled a maximum of $20 million to al-Qaeda every year.* Acting on such information, as well as a belief that al-Barakaat's founder was Ahmed Nur Ali Jimale, a close associate of al-Qaeda, federal agents of the United States carried out raids against four different al-Barakaat storefront operations within the United States.

These raids on one of the major hawala networks frightened many participants throughout the terror financing community, encouraging numerous hawalas to comply with the registration requirements. However, the government's case was soon weakened as it became evident that the action was based on a tenuous connection to terrorism. Thus, the investigation served only to freeze people's primary source of income. Specifically, al-Barakaat operated very similar to a bank, and thus held

---

* The Financial War on Terrorism and the Administration's Implementation of Title III of the USA PATRIOT Act: Hearing before the S. Comm. on Banking, Hous., and Urban Affairs, 107th Cong. 16 (2002) (statement of Kenneth W. Dam, Deputy Secretary, U.S. Department of the Treasury).

individuals' assets. When the U.S. government seized those funds, many individuals were left entirely without access to their livelihoods. Due to the concerns raised by the freezing of assets, the international community protested the "strong arm" tactic used by the United States. This case is directly applicable, though, to the discussions of this chapter, where the question remains how the government can best shut down those who fund terrorism.

While the al-Barakaat case is an example of the U.S. government proactively stepping in to shut down an entity believed to support terrorism, it also illuminates many of the inherent problems. For instance, the government's raid on al-Barakaat, one of the largest hawala networks, occurred only in four U.S. states, presumably permitting the organization to continue its full operations elsewhere in the world. Further, the fractious nature of hawalas shows how mere financial raids and seizures struggle to effectively attain the ultimate goal of curbing terror financing. If one entity is shut down, there will be a new entity on the next block by the end of the week.

In 2002, Operation Green Quest agents arrested Mohamed Albanna, a Yemeni American, on charges of operating a hawala without the requisite licensing. The federal government alleged that Albanna had transferred over $3 million into Yemen since November 2001. Not only did Albanna fail to comply with the aforementioned reporting requirement for a transfer of more than $3,000, but he also neglected to register his business with either the State of New York or the federal government. Similar to al-Barakaat, postarrest evidence showed that the actual ties to terrorism were more tenuous than originally believed by prosecutors, who subsequently conceded that the money in question was not actually used to fund terrorism. Rather, many of those who participated in Albanna's business argue that they were simply following the religious tradition of their heritage by sending money to family back home.

The final PATRIOT Act case study broadens the terror financing question beyond just hawalas. An investigation into the Islamic Saudi Academy, located in Northern Virginia, highlights the fact that questions of terror financing are far broader than just hawalas, but rather apply anywhere that funds further a terrorist agenda. This school, although admittedly less extreme than many madrasas in Muslim states, teaches grammar school and high school students not only regular academics, but also that a day of judgment awaits the conversion of the world to Islam and the elimination of the Jewish people.

The academy came under particular scrutiny when a valedictorian was charged and convicted for participating in plans to assassinate

President Bush. Further, a member of the school's financial board was arrested while videotaping the Chesapeake Bay Bridge. Thus, the U.S. government has begun an investigation into the origins of the school's funding, as well as an inquiry into the specific role played by the Saudi government. The question of the Saudi influence was given some attention in the 9/11 Commission Report, where it was noted that the Saudi government spends funds to disseminate Wahhabi beliefs to mosques and educational institutions around the world.

These case studies highlight the tenuous balance required in the efforts to curb terror financing. First, the above examples show that since 9/11 the U.S. government has been willing to act in a broad and forceful manner toward suspected financers of terrorism. Second, though, these actions have raised great concern among the people "on the ground," who use IVTS networks for completely legitimate purposes. Thus, the recommendations at the conclusion of this discussion must take into account the delicate balance between these often contradictory interests. Furthermore, these case studies highlight the delicate balance existent in every national security and homeland security discussion—balancing between legitimate government needs and legitimate civilian actions. As an example of this balancing, in the above-discussed case studies, there were many complaints that violations of a registration requirement cannot equal or necessitate the draconian act of freezing assets.

However, weighing toward the government's interest, the registration requirements are a minimal and simple way to attempt this delicate balance in light of other possible remedies. While hawalas often serve a legitimate purpose, they are also ripe for the picking by unscrupulous investors. As such, the government merely requires the registration of these businesses, rather than finding them illegal per se. In essence, these cases demonstrate two things. On the one hand, they show governmental power, the use of which is central to the discussions and recommendations of this chapter. On the other hand, these case studies highlight the delicate balance required by the government in regulating financial transactions related to hawalas. The prosecution of these entities, as well as the freezing of assets, should send shockwaves throughout the hawala community, coercing compliance with fairly meager reporting requirements.

However, such governmental actions may also serve to embolden the unscrupulous investors and hawaladars. Clearly, the government could avoid the problems highlighted in the al-Barakaat example by employing a team of analysts to fully engage and analyze the entity in question, but such deference to the entity would lead to multiyear government investigations, a reality that is not practical in the world of counterterrorism. Not only is the U.S. government engaged in a new and rapidly changing struggle against terrorism that precludes the allotment of such large amounts of time, but the hawalas also do not lend themselves to such inquiries due to a complete lack of records.

The final lesson from these case studies is the recurring theme of a need for international cooperation. If money laundering and terror financing are treated merely as domestic crimes, the goal of eliminating terror financing cannot be achieved. Rather, the government must proactively establish and exercise universal jurisdiction over this international crime.

The fact that terror financing takes place in more calm settings than warfare makes it no less of a threat than an individual with a bomb. Conversely, financiers should receive greater attention than the foot soldiers, as there are a thousand people willing to wear a bomb. Killing one bomber only makes room for the next person to pick up the bomb. However, the financiers are few and far between. Thus, taking these individuals out of circulation creates a larger dent in the furtherance of the terrorist goals. As the "zone of combat" in the "war on terror" is continually expanding beyond traditional notions of warfare, terror financing needs to be included as part and parcel of this expanding zone of combat.

FOREIGN TREATMENT OF TERROR FINANCING

As has been articulated above, domestic efforts to curb terror financing must simultaneously take an international perspective. Thus, it is important not only to discuss the specific international treatment of terror financing, as discussed earlier, but also to discuss the current steps being taken in foreign jurisdictions—such as Israel and England—to curb terror financing.

Israel

A starting point for discussing Israel's response to terror financing is the Prohibition of Financing Terrorism Law of 2004 (PFTL). The PFTL was passed in direct response to the United Nations International Convention for the Suppression of Financing Terrorism, a resolution specifically defining terror financing as an offense and encouraging nations to enact legislative measures intended to identify, locate, and seize funds intended for terror financing.

Although this international convention was passed in 1999, well before the 9/11 attacks, it was not until after the attacks that the United Nations Security Council enacted Resolution 1373, specifically calling on states to work jointly against terror financing by complying with the convention. Specifically, Resolution 1373 created an international obligation for nations to criminalize activities related to terror financing, criminalize possessing assets on the behalf of others connected to terrorism, and allow for the freezing of assets known to be tied to terrorism. Further, Resolution 1373 created the Counterterrorism Committee, a body assigned the task of watching over the implementation of the resolution in all states. The general goal of this resolution, and of the committee charged with its oversight, is to broaden the world's efforts to combat terrorism by requiring all states to participate in preventing terror financing.

The Israeli government's passing of the PFTL highlights the international response needed in combating terror financing. Terror financing is not an offense that can be handled solely through existing domestic frameworks. Rather, Israel realized the need to subjugate domestic law to an international standard. Thus, the PFTL provides Israeli authorities greater strength in combating terror financing, as they must now act under international standards in addition to domestic standards.

For example, Israeli authorities have established provisions for criminalizing the act of rewarding a terrorist action, classifying an entity as an international terrorist organization with no connection to Israel, handling an Israeli citizen who finds himself involved in a monetary transaction with suspected terrorist connections, and—the strongest provision instituted by the international standards—permitting the Israel government to administratively seize money with suspected ties to terrorism.

England

The United Kingdom has extensive experience with terror financing, in part because of the troubles in Northern Ireland. In response to the IRA, the UK government worked to cut off terror financing by focusing legislation on robberies and tax fraud. However, as the IRA moved into the political sphere, it began to use more sophisticated networks of organized crime. Thus, the United Kingdom's focus turned toward drug money in the effort to curb terror financing. The events of 9/11 did not radically alter UK legislative efforts against terror financing; rather, the United Kingdom merely tweaked the existing standards in an effort to more effectively stop terror financing.

The first major antiterrorism legislation in the United Kingdom was the Terrorism Act of 2000. However, despite this enactment, the attacks of 9/11 brought English legislators back to the table to further expand counterterrorism law in the United Kingdom. This led to the Antiterrorism Crime and Security Act, which established the power of the government to confiscate money believed to have ties to terrorist organizations. This legislation operates as a check by stripping the judiciary of jurisdiction regarding the freezing of assets located outside of the United Kingdom.

Beyond broadening the government's power to seize assets, the legislation also made it a criminal offense to operate in the "regulated sector" and not inform law enforcement of suspicious activity. In short, the response from the United Kingdom to the escalation of international terrorism follows very closely the legal developments within the United States. The British government acted to ensure that (1) people putting money into the financial system were regulated and that (2) people working in the financial system were also regulated.

The brief discussion of terror financing legislation in Israel and England highlights the fact that (1) this is not merely an issue for the United States to consider, and (2) these international actors need to pool their resources in order to cast an international net to fight terror financing.

FREEDOM OF RELIGION

If questions of homeland security and national security could take place in a vacuum, this nation would be a far safer place and the laws would be far more easily applied. However, no such vacuum exists. Rather, for every homeland security policy determination, the policy maker must grapple with many tangential topics. In this instance, such issues are best shown by discussing the balance that must be made between legitimate homeland security questions and the impact policies can have on the freedom of religion.

Specifically, the question here is the interplay between the governmental eradication of terror financing and the protection of one's freedom to exercise religion. Since 9/11, the United States has been willing to act in a heavy-handed manner in the name of national security, and the Muslim community often suffers some of the hardest blows. However, many see the role of Islam in the 9/11 attacks as justifying such governmental actions. It is important to note, though, that everything done in the halls of the American government establishes precedent. What is done today affecting the Islamic community could just as easily be turned against Christian or Jewish communities in the future. The homeland security policy maker must stay focused on the great need for targeting terror financing while still preserving the freedom of religion.

When developing the Bill of Rights, the framers referred to the freedom of religion as the first freedom. As pointed out by Grace Smith of the Heritage Foundation,* this right was listed first because:

> Religious freedom is a natural right that cannot justly be withheld. Its importance is underscored in the Second Vatican Council's declaration on religious liberty, Dignitatis Humanae, which recently celebrated its 40th anniversary. Promulgated by Pope Paul VI on Dec. 7, 1965, the document reasserts the Catholic Church's teaching that religious freedom is a right that innately belongs to every individual simply because of his or her humanness.
>
> "The right of man to religious freedom has its foundation in the dignity of the person," it reads, "whose exigencies have come to be ... fully known to human reason through centuries of experience." Here is a religious claim about human dignity that can resonate with Catholics and non-Catholics alike in its appeal to reason, experience and moral intuition.

Delving further into history to understand the contours of America's freedom of religion, Congressman James Madison of Virginia and Congressman Fischer Ames of Massachusetts were both instrumental in the adoption of this clause. These two individuals highlight two central reasons for having a freedom of religion clause in the United States. First, Massachusetts still had churches in 1790 that were established by state law. Thus, Congressman Ames noted the problems that would arise if the Federal Congress ever decided to establish a federal church. This foresight led to Ames demanding the prohibition of Congress to institute religion. Second, then, was concern for unity among the American states. There were clearly differences in religious and theological beliefs among the states (as noted by the presence of Congregationalists, Baptists, Quakers, Methodists, and Episcopalians). Thus, instituting freedom of religion would serve to prevent conflicts from trying to coalesce all of these religious vantage points into one cohesive church.

When deciding how homeland security policy can address terrorism financing, questions arise where the policy maker must apply his definitions to the practical world. In this example, such queries include: Is it a

---

* Grace V. Smith, Forum: Rethinking the First Freedom, Washington Times, March 12, 2006, at B5

violation of one's freedom of religion to require IVTS networks to register? Is it a violation of one's freedom of religion to require banks to "know their customers"? Is it a violation of one's freedom of religion to require people giving money through an IVTS to provide identification and maintain a paper trail? The answer to all of these should be no. The government must be able to enact legislation in the name of counterterrorism, and as terror financing is at issue in such an effort, legislation must be permitted to impact the financial markets and transfer systems without an overly burdensome concern for a tangential impact on religious freedom. But, only with a thorough and consistent understanding of what is and what is not proper homeland security policy can a policy maker analyze such questions.

At the outset, it is incorrect to suggest that freedoms are absolute. This interpretation would become cognizable to anyone who chooses to go into a crowded movie theater and yell "fire." Despite a right to free speech, there are limitations, one being that you cannot yell "fire" in a crowded area without the existence of an actual fire. Thus, this discussion must begin with an understanding that even though we hold our rights closely in the United States, they are all subject to some level of limitation, the question merely being how much limitation.

In balancing national security and the freedom of religion in the context of regulating the financial sector, the operative question concerns the centrality of giving money through IVTS networks to Islam. This is the central question because the more fundamental the use of IVTS networks is to Islam, the greater the impact regulations may have on religious freedom. The Specifically Designated Terrorist (SDT) list, created by President Clinton's Executive Order 12,947, serves as an effective analytical tool. This list was then broadened by President Bush after the 9/11 attacks through Executive Order 13,224, which created the Specially Designated Global Terrorist (SDGT) blacklist. This second list broadens Clinton's list through the inclusion of global terrorists, rather than only groups participating in the Middle East peace process. The creation of both lists raised multiple freedom of religion claims from charities on the blacklists.

To examine the question of whether the government's legislative action against SDGTs comports with the Constitution's freedom of religion, a discussion of the Holy Land Foundation (HLF) case is illuminative. HLF argued, similar to many other SDGT organizations, that the blacklist was in violation of the Religious Freedom Restoration Act, and thus was a violation of the freedom of religion in general.

In *Holy Land Foundation for Relief and Development v. Ashcroft*, the district court for the District of Columbia employed two lines of analysis. The court first analyzed HLF's argument of substantial burden on its free exercise of religion, and second, the court analyzed HLF's argument of substantial burden on behalf of HLF's donors and employees. In short, HLF argued that its work in accepting and using donations from Muslims was the free exercise of religion, as such actions fulfilled the Islamic religious obligation of zakat.

The court responded by holding that HLF failed to show that it was, in fact, a religious organization rather than just a "nonprofit charitable corporation." Thus, if an entity is found not to be a religious organization, it follows that there is no need to reach a discussion of RFRA. The second holding, then, is not applicable here, as the court rested its denial on a question of the standing of HLF to raise an argument on behalf of third parties. Applicable here, though, is the dicta from the D.C. Circuit where the court held that HLF was participating in the furtherance of terrorism, and as terrorism is not mandated by any religion, promoting terrorism is not protected. Thus, this case shows that legislation prohibiting the financing of terrorism cannot in and of itself be argued as a violation of the freedom of religion clause because the financing of terrorism cannot be argued to be an exercise of religion. However, such a broad statement may fail to recognize that legislation aimed at terrorism has tangential impact that infringes on a particular right. As such, the operative issue remains: balancing.

Although this discussion highlights the proposition that the government possesses the power to enact legislation aimed at the institutions purportedly financing terrorism, the corollary question is whether the government can do so if such actions impact the donor. Similar to the judicial response to the IVTSs, which would be the same as the response to HLF's arguments, an individual donor must show that his or her act of giving money to an IVTS is a religious act, that the entity receiving the money is a religious organization, and that the legislation unduly infringes on the use of the IVTS. Donors would have to emphasize that these entities and charities exist not merely for their humanitarian work, but rather for their direct role in providing a place for Muslims to practice zakat.

However, the SDGT list and other financial actions against purported financiers of terrorism merely prohibit the use of charities and monies for terrorism. Thus, unless terrorism is a religious mandate, then both charities and donors lack standing to raise such arguments. Implicit, though, in this balance is the fact that the discussion is not truly this black and white. The IVTS may, in fact, be used as an avenue for religious exercise, and the antiterror financing legislation may only be targeted at giving money to terrorism. But an individual may feel unduly spied upon by any antiterror financing legislation despite the person's wholly innocent use of the IVTS network.

RECOMMENDATIONS

Terrorism financing highlights the great need for a clear understanding of what is and what is not homeland security policy. As we are likely to grant the government at least some modicum of deference when it purports to act in the name of homeland security, the risk that such actions will negatively impact such previous rights of religious freedom requires that we operate with a very clear understanding of what is actually homeland security policy. The government of the United States, through the Department of Justice and Department of Homeland Security, has a duty to act in a proactive way to eliminate terrorism. In doing so through a proper balancing of all interests, any recommendations must respect: (1) the freedom of religion, (2) the free market, and (3) the need for national security.

With these considerations in mind, the approach is twofold. First, the government bears responsibility to guarantee that its actions are both effective and not overly broad—consistent with a clear understanding of homeland security. Second, from the grassroots, "man on the street" perspective, there are enormous responsibilities in making sure that people will not abuse the IVTS networks in support of terrorism. To this end, the following recommendations aim at both the top-down governmental role and the bottom-up individual role in attempting to orchestrate a system that will both curb the financing of terrorism and protect one's right to free exercise of religion:

1. When the government freezes the assets of an entity, the government must proactively look through the registered transactions to decipher which of them did not support terrorism. Those transactions that are then found not to support terrorism must be restored as quickly as possible.
2. A record of all transfers made through any form of an IVTS must be kept by both the IVTS agent and the consumer transferring the money. This will provide an easier window

133

through which to view legality, making it far more likely that the government will not need to broadly shut down organizations due to a lack of information.

3. In the event that an IVTS is suspicious that its system is being used for the illicit purposes of financing terrorism, the IVTS agent must both refuse to effectuate such a transfer and notify the proper governmental authorities.

4. The government must be able to articulate a *prima facie* case for any action that either shuts down an IVTS or freezes the assets of such an entity. This requirement will serve to prevent the government from overstepping its authority in sweeping actions.

5. The U.S. government must take a proactive approach to IVTSs and hawalas where the government investigates the existence thereof, rather than tacitly using reporting requirements. Under this proactive approach, the government would have an enforcement wing whose sole duty is to find IVTS networks existing on American streets. This is not to proactively shut these entities down; rather, the job is to only make a record of each network's existence, so as to make monitoring easier.

6. Traditional concepts of money laundering cannot be used in fighting terrorist financing; rather, the government must look for any illicit intent, not whether the money itself is clean or dirty at the time of the transfer. Transferring clean money must make the money illegal immediately upon either a showing of a future intent to support terrorism or a willful blindness to such support.

7. Any efforts to curb the financing of terrorism must take significant steps to proactively protect the freedom of religious exercise. When establishing limits and procedures for IVTS networks, such procedures cannot be so burdensome that they infringe on one's freedom to practice one's religion. But, from the ground level, individuals using IVTS networks must recognize the abuses being perpetrated upon these networks, and in acknowledging those abuses, they must accept a heightened amount of scrutiny.

This discussion employed the essential tool of balancing. The act of balancing when formulating effective antiterrorism financing legislation considers the government's interest in protecting the citizenry and the competing interest of the individual's freedom to freely exercise religion. Throughout this morass of interests, the concluding recommendations highlight that the answers are neither purely governmental nor individual. Rather, government and public alike must assume their responsibilities to effectuate an end to the abuse of financial markets in the name of financing terrorism.

8

Business Continuity

In the questionnaire distributed to business leaders I asked two questions. (1) How does your corporation assess threats/risks/dangers? (2) How does your corporation prioritize threats/risks/dangers? While questionnaire participants had much to say on the rest of the questionnaire, they remained tellingly silent on these two questions. Many were not sure, or wrote "not applicable." These answers suggest business leaders have not had the critical discussion of what threats they face, and more importantly, how to respond. Needless to say, this is a failing of extraordinary import with both short-term and long-term ramifications affecting both the specific enterprise and the larger community.

However, in numerous conversations with business leaders I was reassured (perhaps they were reassuring themselves) that this was an issue they intended to address in the near future. That consistently struck me as an inherently odd response: If you (as a business leader) know there are potential threats with the ability to affect your assets (physical and human alike), does corporate responsibility not require implementing a sophisticated business continuity plan subjected to simulation exercises ensuring effectiveness and resilience? The easy—and frankly—obvious answer should be a resounding yes. The response I have consistently heard is: "Indeed, this is something we need to address." That is not to suggest all corporations—regardless of size—are lagging in this vein; it is, however, to articulate a deep concern: many are failing to act proactively and responsibly. The potential fallout is enormous.

U.S. military engagement in Iraq and Afghanistan has a definite impact on American corporations, whether their assets are based in the United States or worldwide. To that end, American corporations must develop, articulate, and implement geopolitical sophistication that facilitates minimizing asset vulnerability while maximizing profitability in a combustible environment. Business continuity—this chapter's theme—is essential to this process; simply put, businesses must invest significant resources in developing viable and effective business continuity models. This must be done—as the old adage says—sooner rather than later.

As the oil spill in the Gulf of Mexico dramatically illustrates, planning for a major event requires considering not only technical issues relevant to a particular industry; as evidenced by the dismissal of BP CEO Tony Hayward the failure to develop plans addressing the human toll of both a crisis and how to effectively communicate with multiple constituencies and audiences has significant consequences. Simply stated, BP is an effective case study in how **not** to prepare and respond to a major homeland security event. From failing to sufficiently prepare "what ifs" to ignoring concerns previously raised regarding equipment and infrastructure to failing to develop a plan facilitating protection of potentially affected economic sectors in the local community (to communication messages that were, at best, embarrassing and, at worst, shocking), BP reflects the importance of business continuity from the perspective of numerous audiences, including:

1. The specific company/enterprise
2. The local community—economically, morally, and socially
3. Shareholders
4. The larger community, domestic and international alike (circumstances dependent)
5. Company employees and their families (especially if employees are injured or killed)
6. Elected officials (congressional hearings and a meeting with President Obama reflect the impact of not having a plan in place and a failure to compellingly articulate mitigation efforts)

When creating business continuity plans, corporate leaders must understand that quarantine, evacuation, risk assessment, resource allocation, cost–benefit analysis, and legal liability are but a handful of questions that must be addressed. Undertaking this effort means that potential points of vulnerability must be identified in order to determine how best to protect corporate assets, defined as people and property alike. Corporate leaders must develop contingency plans; these enable the corporation to continue its activity. This can only be accomplished by simultaneously addressing and providing for basic needs during and after a major incident and implementing a previously developed long-term recovery plan. Together, the two enable, or at least facilitate, continuity in the face of a major event, be it terrorism or natural disaster. Doing this proactively does not ensure that a major incident will not negatively affect the enterprise. It does, however, create an infrastructure whereby the fallout of a particular incident can be minimized and mitigated.

To make the point: Ensuring continuity requires addressing basics, including food, water, medical attention, and enabling employees to communicate with their families. In an effort to minimize the impact of a major homeland security event, it is incumbent upon corporations to consider—carefully and strategically—how to most effectively protect their assets; this includes the safety of their employees. In doing so, corporate leaders must inherently understand that not all assets can be equally and uniformly protected. By analogy, superimposing the 18 critical infrastructures tasked by DHS to a corporation facilitates understanding both the practical essence of business continuity and its possible, actual implementation. That is, by carefully analyzing the applicability and relevance of the 18 infrastructures below to their specific industry, corporate leaders can more effectively develop a business continuity model appropriate to their industry. Simply put, the 18 tasked critical infrastructures can serve as an effective checklist to business leaders developing a business continuity model.

By example: Business continuity requires planning for how to ensure efficient and reliable food supply to employees in the event of quarantine; similarly, a major event demands fully functioning alternative communication infrastructure and facilities and the availability of sophisticated, emergency healthcare and public health resources.

1. Agriculture and food
2. Banking and finance
3. Chemical
4. Commercial facilities
5. Communications
6. Critical manufacturing

7. Dams
8. Defense industrial bases
9. Emergency services
10. Energy
11. Government facilities
12. Healthcare and public care
13. Information technology
14. National monuments and icons
15. Nuclear reactors, materials, and waste
16. Postal and shipping
17. Transportation systems
18. Water

Corporate leaders must look their employees in the proverbial eye and be straight shooters. This is the only way to prepare them for emergency measures, including extended lockdown. There is a need to call it what it is. In a bioterrorism attack, people may become carriers of agents; as a carrier, they potentially endanger thousands, including their own children. That is the reality for which corporations must prepare. To refer to *quarantine* as the "q word," as if we fear it, as at a recent mini bioterrorism simulation, irresponsibly minimizes its significance.

Insufficient attention to the three Ps (preparation, planning, and practice) leaves corporations vulnerable to significant legal liability. Corporate leaders are under enormous pressure to ensure tomorrow's earnings. In the aftermath of 9/11 and Katrina, they need be similarly concerned with tomorrow's attack. Preparations for today and tomorrow are a corporate leader's primary responsibility. Responsibilities may be delegated, but the CEO is the "captain of the ship."

Business continuity—if properly implemented—facilitates the continued functioning of a corporation in the immediate aftermath of either a natural disaster or a terrorist attack. It is, in many ways, a proactive insurance policy intended to facilitate corporate functioning in the face of a significant event. Uncertainty and panic are predictable responses. Their impacts, short term and long term alike, can be minimized by sophisticated planning. The required planning must be predicated on practical considerations; however, to be effective, the preparation must reflect *both* tactical and strategic thinking.

Business continuity requires corporate leaders to honestly assess—and answer—the question: "Where are we most vulnerable?" Furthermore, to develop an effective and realistic business continuity model requires a thorough understanding of one's own supply chain. The beans-to-cup approach discussed in Chapter 2 facilitates an effective and efficient business continuity model because it segments the production process into identifiable and definable parts.

Segmenting enables corporate leaders to develop business continuity models that facilitate responses to damage (natural disaster or terrorism) to each segment, both individually and collectively. That is, each segment can be viewed both as a silo (stand alone) and as part of an integrated whole. In developing a business continuity model, corporate leaders must prioritize the value of each segment in an effort to determine whether to include that particular segment—in the event of an attack or disaster—in the corporation's business continuity model. That is, corporate leaders must determine the relative importance of each segment in the supply chain, both as a single entity and as part of the larger enterprise.

In the context of homeland security, decision making in a crisis is extraordinarily difficult because it requires anticipating and assessing possible threats to society at large and the corporation in particular. This is not a trivial exercise because it demands both tough financial decisions and understanding that—in practical terms—different assets will be protected to different degrees; some segments will not be the recipients of full protection and dedicated rehabilitation.

This requires immediate implementation of the three Ps. That is, in addition to realization, regarding the absolute requirement to develop business continuity models, corporate leaders must ask the *how* question. Recognition or identification of the requirement is essential; the requisite measure is to articulate, develop, and implement a plan that facilitates business continuity. However, in addressing business continuity and developing simulation exercises (of crucial importance), corporate leaders must determine their definition of the term, what are their goals, and in what time frame.

In the ideal, the entity will be fully functioning in no time in the aftermath of an attack; in reality, rebounding is dependent on an extraordinarily broad range of factors, some well beyond the control of management. Business continuity does not happen in a silo or vacuum; while a specific entity may be the victim of an attack (i.e., animal rights activists targeting a drug company that conducts experiments on animals), law enforcement's capability to respond is dependent on external factors predicated on additional circumstances. Conversely, if the attack (on a corporation) is but one in the context of a broader attack, then law enforcement's abilities are even more minimized, with already thin resources further stretched.

In developing business continuity models I have emphasized to corporate leaders that they *must* understand broader geopolitical considerations and ramifications. While they are, understandably, focused on their enterprise (for which they have legal responsibility), the larger community (public, law enforcement) has additional factors to weigh in determining its response to a particular attack. While that places the onus on corporate leadership with respect to their entity, it also requires them to understand—perhaps begrudgingly—that other entities also deserve the full attention of decision makers and law enforcement. This reality reinforces the requirement that corporate simulation exercises include external organizations (local government, law enforcement, media). A business continuity model that incorporates internal and external entities alike is far more effective—and efficient—than a singular (actually, insular) model.

This is, of course, easier to suggest than implement, primarily because of concern of vulnerability to corporate piracy with respect to protection of intellectual property. Corporate leaders are justifiably concerned that business continuity models that include external entities potentially enhance their vulnerability. While that concern is justified, a balancing analysis suggests that inclusion of others in the corporation's business continuity model and simulation exercises has significant long-term, strategic benefits.

By way of example: A major American corporation with significant international assets conducted—at the CEO's initiative—an in-house bioterrorism simulation exercise that included quarantining company employees for a number of days. The exercise based on the three Ps was intended to enhance management and employee understanding of threats faced by the corporation and to simulate responses (tactical and strategic) facilitating business continuity. There is little doubt that *not* implementing the three Ps will significantly affect corporate response time in the face of a real attack; that is clear from weaknesses discovered in simulated attacks. As part of the hypothetical attack, the corporation imposed quarantine on its employees; without doubt, this is one of the most controversial aspects of short-term business continuity. In that vein, the results were less than satisfactory: plans did not provide for backup generators, food, or water for the quarantined employees. This is similar to what was demonstrated in the nation's capital December 16, 2010 when an alleged shooting led to a lockdown of congressional offices who were left to fend for themselves, with minimal supplies of food or drinks for those locked down.

Simply stated: Quarantine, if rigorously applied, denies employees the right to leave the workplace. It obviously raises significant legal questions; the employer, in essence, is deputized to deny the employee basic freedom of movement. Quarantine is of particular importance in the business continuity model for two reasons:

1. It enables the employer to expeditiously begin recovery (the employee is forced to be on-site).
2. It maximizes—or at least facilitates—continued functioning *during* the attack (as distinct from continuity *after* the attack).

Conducting a response exercise is critical, and for this, the corporation is to be lauded. The responses simulated addressed a broad range of issues, including:

1. Where are designated meeting points in the event of an attack?
2. What employees are tasked with leadership roles in the event of an attack?
3. How are employees to be accounted for in the event of an attack?
4. What will be information sources during an attack?
5. How are secondary assets (off-site from principal corporate headquarters) protected during an attack on the principal asset?
6. Are employees to be quarantined?
7. Are there sufficient provisions on-site in the event of a quarantine?
8. What lines of communication have been established with law enforcement and other public officials (preferable that local law enforcement officials participate in the exercise)?
9. What officials are designated company spokesmen with respect to media (whether broadly or narrowly defined) and employee family members?
10. How has the company defined successful and effective continuity? Have corporate leaders created a goals predicated matrix?

While business continuity is forward looking, it must be based on an assessment of present needs and future threats; just as incorporating external forces is recommended (thereby ensuring the planning is not silo predicated), the planning must also reflect sophisticated understanding of the particular industry and its vulnerabilities. See the following section by way of example.

CHEMICAL FACILITIES

Chemical facilities may be an attractive target for terrorists intent on causing economic harm and substantial loss of life. Precisely for that reason, chemical plant facilities are highly relevant to a discussion regarding business continuity. While a CEO's decision to ignore viable threats to his particular asset/facility is troubling in its short-sightedness and irresponsibility, it does not necessarily affect a broader community. That is not the case with the chemical facilities, for an event—whether terrorism or otherwise—has the potential to cause extraordinary harm with respect to injury and death. That is, not only is the potential economic fallout significant, but the possible human cost is literally unimaginable.

Some suggest applying the chemical plant paradigm in the context of a business continuity discussion is akin to fearmongering, primarily because the danger inherent is fundamentally distinct from threats posed to the majority of U.S. businesses. That is both correct and incorrect. It is correct because indeed the danger posed by an attack on a chemical plant is fundamentally distinguishable from an attack on a clothing store.

Simply put: An attack on a clothing store may harm those either inside or nearby; an attack on a chemical plant has enormous significance that goes well beyond those physically at or near the plant. Conversely, it is incorrect because the failure to develop a business continuity model for a chemical plant unnecessarily and irresponsibly places people and assets at risk; the exact same holds true for where the reader of these lines bought coffee and pants yesterday. While perhaps convenient, for business leaders to assume their facility is invulnerable or that their employees will know how to react in a crisis shows the extraordinary fallacy of this common assumption.

In discussing business continuity models with business leaders—in the United States, Europe, and Israel—I consistently draw on the chemical plant industry. I do so knowing it is viewed as an exaggeration and largely inapplicable to that specific leader's particular industry. However, when I explain

144

the universality and commonality of the threat faced in the event of inadequate planning, dubious looks reflecting skepticism are generally (never say always) replaced with cautious nods of understanding. Here's why:

1. The realization that no plan is in place is sobering.
2. The realization that even secondary targets (for example, businesses in the World Trade Center) are highly vulnerable and that the primary target (for example, the World Trade Center) offers, at best, minimal protection.
3. The domino theory: If x is vulnerable, then my asset is also potentially vulnerable.
4. The realization that if the chemical plant industry is largely unprotected—in spite of its obvious vulnerability and importance—then assets otherwise assumed protected are similarly at risk.

Many facilities exist in populated areas where a chemical release could threaten thousands of lives. For example, a CBS News report in June 2004 reported: "There are more than 100 chemical plants—in backyards all across the United States—where a catastrophic accident or an act of sabotage by terrorists could endanger more than a million people. One plant in Chicago could affect almost three million people; in California, the chemicals at one site have the potential to kill, injure, or displace more than eight million people."[*]

A 1998 report, *Too Close to Home: Chemical Accident Risks in the United States*, by the U.S. Public Interest Research Group and the National Environmental Law Center, addressed the distribution of chemical facilities in the United States relative to population centers.[†] The report stated that "more than 41 million Americans live within range of a toxic cloud that could result from a chemical accident at a facility located in their home zip code." A terrorist attack, not simply an accident, can result in the release of such toxic chemicals.

---

[*] U.S. Plants: Open to Terrorists, CBS News (June 13, 2004), http://www.cbsnews.com/stories/2003/11/13/60minutes/printable583528.shtml (last accessed December 24, 2010).

[†] Allison Laplante, Too Close to Home: A Report on Chemical Accident Risks in the United States, U.S. Public Interest Research Group http://uspirg.org/uspirg.asp?id2=5067&id3=USPIRG&

FIGURE 8.1   A nuclear or chemical attack by terrorists is a real and credible threat.

The threat of a terrorist attack against a chemical plant is real and it is credible (Figure 8.1). The Department of Homeland Security has issued warnings of potential attacks on chemical facilities.* The Department of Justice has also concluded that the risk of terrorists attempting to cause an industrial chemical release in the foreseeable future is both real and credible.† The chemical industry has also recognized the risk of a terrorist attack on a chemical manufacturing or storage facility. The American Chemistry Council has released numerous statements calling

---

* James L. Beebe, Inherently Safer Technology: The Cure for Chemical Plants Which Are Dangerous by Design, 28 Hous. J. Int'l L. 238, 243 (2006).
† Id. at 244.

for federal legislation to address security at chemical facilities.* The current American Chemistry Council president and CEO, Jack N. Gerard, recently stated:

> Four years is simply too long to wait following the terrorist attacks of 9/11 for Congress to help safeguard the men and women who make our nation's essential chemical products. America needs a comprehensive, federal plan to secure the critical chemical infrastructure that every sector of our economy relies on. We have testified before Congress in favor of tough federal legislation to ensure that all companies across America that make or handle chemicals are doing everything feasible to counter terrorism.[†]

Government reports and the chemical industry trade associations make it clear that the threat of a terrorist attack against a chemical facility is real, and that some form of federally mandated fixed-site security requirements are necessary. Currently, however, there are no federal laws that explicitly require chemical facilities to assess vulnerabilities or to take security actions to safeguard their facilities from terrorist attacks.[‡]

There are two key federal laws that require or encourage certain chemical facilities to reduce risks to the general public from releases of hazardous chemicals: the Emergency Response and Community Right-to-Know Act (ERCRA) and the Clean Air Act (CAA).[§] These laws are principally environmental laws and do not explicitly address the concerns associated with a terrorist attack. The legislation does, though, have some bearing on the conduct of chemical facility operators and how they handle and safeguard hazardous substances.

Local emergency response committees (LERCs) coordinate planning and response to potentially large releases of "extremely hazardous substances."[¶] Chemical facility managers are required to provide informa-

---

* See generally Martin J. Durbin, Managing Director of Security and Operations for the American Chemistry Council, Chemistry Facility Security: What Is the Appropriate Federal Role? Testimony before the Senate Committee on Homeland Security and Governmental Affairs (July 13, 2005).

† American Chemistry Council Addresses Proposed Security Legislation, SecurityInfoWatch.com, http://www.securityinfowatch.com/article/printer.jsp?id=6813 (last accessed November 19, 2010).

‡ John B. Stephenson, Federal Action Needed to Address Security Challenges at Chemical Facilities, U.S. General Accounting Office, Testimony before the Subcommittee on National Security, Emerging Threats, and International Relations, Committee on Government Reform, House of Representatives, GAO-04-482T (February 2004).

§ Linda-Jo Schierow, Chemical Facility Security, CRS Report for Congress (January 12, 2006), 15.

¶ Id.

tion to LERCs and local emergency responders about chemicals present at the facilities and to notify those officials in the event of a sudden release of any of the listed chemicals.* The law stops short of mandating the facilities to assess the risks that may cause the release of the extremely hazardous substances, nor does the law require the chemical facilities to take defined, deliberate steps to reduce the risk of chemical releases.†

The CAA §112(r) imposes a general duty on chemical facilities producing, processing, handling, or storing any "extremely hazardous substance" to detect and prevent or minimize *accidental* releases and to provide prompt emergency response to a chemical release in order to protect human health and the environment.‡ The CAA requires owners and operators of chemical facilities to prepare risk management plans (RMPs) that summarize the potential threat of sudden, large releases of certain chemicals.§ The CAA also requires chemical facilities to have plans in place to prevent chemical releases and to mitigate any damage caused by a chemical release. The Environmental Protection Agency (EPA) was given supervisory responsibility.¶ The EPA is required to review RMPs regularly and, if necessary, require revisions if the plans are inadequate.** Chemical facilities also have reporting requirements under CAA §112(r). Chemical facilities are required to report their RMPs to the EPA and other governmental agencies.†† In fact, the Department of Justice (DOJ) was directed to report to Congress on the extent to which RMP regulations led to actions that are effective in detecting, preventing, and minimizing the consequences of releases of regulated substances that may be caused by criminal activity, the vulnerability of facilities to criminal and terrorist activity, and current industry practices regarding site security.‡‡ The DOJ reporting deadline was August 2002 and the DOJ failed to make this deadline.§§ The DOJ congressional report is still pending.

---

* Id.
† Id.
‡ Id. at 16.
§ Id.
¶ Id. at 23. Homeland Security Presidential Directive 7 rescinds the EPA supervisory role. The authority for overseeing the security of chemical facilities has been transferred to the Department of Homeland Security (DHS).
** Id. at 17.
†† Id. 16–20.
‡‡ Id. at 20.
§§ Id.

Homeland Security Presidential Directive (HSPD) 7 transferred all of the EPA authority for overseeing the security of chemical facilities to the Department of Homeland Security (DHS). The directive requires the DHS to "identify, prioritize, and coordinate the protection of critical infrastructure and key resources that could be exploited to cause catastrophic health effects or mass casualties."[*] HSPD 7 also requires DHS to conduct or facilitate vulnerability assessments of the chemical sector and encourage risk management strategies to protect against and mitigate the effects of attacks on chemical facilities.[†]

Two additional congressional acts impact certain chemical facilities. The Public Health Security and Bioterrorism Preparedness Act of 2002 requires community water systems to perform vulnerability assessments and to prepare emergency preparedness and response plans.[‡] This law is applicable to the chemical industry because some water facilities handle large quantities of chemicals and may qualify as a chemical facility by the chemical industry. The other law that has had an impact on chemical facilities is the Maritime Transportation Security Act (MTSA).[§] The MTSA requires the DHS secretary to identify port facilities (including chemical facilities located in ports or in close proximity to ports[¶]) that pose a high risk of being involved in a transportation security incident and to conduct a vulnerability assessment of such facilities.[**] The facility owners covered under MTSA are required to submit security plans and incident response plans to DHS.[††]

The existing legislation does not explicitly address chemical facility security. The legislation and the presidential directive provide some oversight and general guidelines, but fail to offer any definitive guidance concerning security requirements at chemical facilities.

---

[*] Id. at 23.

[†] Id.

[‡] P.L. No. 107-188, 116 Stat. 594 (2002).

[§] P.L. No. 107-295, 116 Stat. 2064 (2002).

[¶] See Protection of Chemical and Water Infrastructure, U.S. General Accounting Office, Report to the Honorable Robert C. Byrd, Ranking Member, Subcommittee on Homeland Security, Committee on Appropriations, U.S. Senate, GAO-05-327 (March 2005), at p. 4, www.gao.gov/cgi-bin/getrpt?GAO-05-327 (last accessed November 23, 2010).

[**] Id. at §102.

[††] Id.

PRIVATE SECTOR INITIATIVES

The two principal trade associations in the chemistry industry have adopted a code of self-regulation with regards to chemical facility security in lieu of the nonexistent federal legislation. The American Chemistry Council (ACC) and the Synthetic Organic Chemical Manufacturers Association (SOCMA) adopted the Responsible Care® Security Code in 2002. The Responsible Care Security Code requires all trade association member companies to perform vulnerability assessments, develop plans to mitigate vulnerabilities, take actions to implement the plans, and undergo a third-party verification that facilities implemented the physical security enhancements.* There are about 2,300 ACC and SOCMA chemical manufacturing facilities following the Responsible Care Security Code, 1,100 of which are among the estimated 4,000 chemical manufacturing facilities identified by the EPA that pose the greatest risk to human health and the environment if an event were to cause a chemical release at their facility.† Thus, there are still a large number of chemical facilities that may not necessarily follow a designated system of acceptable security measures.

The seemingly large number of chemical plant facilities that do not adhere to an industry or governmental standard for security requirements or safeguards against terrorist attacks has prompted the chemical industry trade associations to push Congress for federally mandated chemical facility security requirements. The principal argument is the need to safeguard Americans and to safeguard the environment. The other argument for federal legislation is that the economic playing field in the chemical industry needs to be level. All major chemical facilities should have to make the necessary monetary expenditures to ensure that the volatile material on their site is properly safeguarded. The industry stresses that this requirement should apply to all major chemical facilities, not just those that voluntarily choose to implement security plans.‡

---

* The principles of the Responsible Care Security Code are listed on the ACC website at http://www.americanchemistry.com/s_acc/bin.asp?CID=373&DID=1255&DOC=FILE. PDF" (last accessed November 25, 2010).
† Protection of Chemical and Water Infrastructure, U.S. General Accounting Office, Report to the Honorable Robert C. Byrd, Ranking Member, Subcommittee on Homeland Security, Committee on Appropriations, U.S. Senate, GAO-05-327 (March 2005), at p. 6, www.gao. gov/cgi-bin/getrpt?GAO-05-327 (last accessed May 23, 2006).
‡ Id. at p. 7.

PRIVATE-PUBLIC COOPERATION

Public-private sector cooperation must extend beyond terrorism; after all, homeland security encompasses a wide range of domestic threats, including natural disasters. Post-9/11 and in the wake of Hurricane Katrina, one of the most important lessons learned by the United States was the dire consequences of the breakdown in communications of governmental agencies, among themselves and with the private sector. Ineffective communication directly led to hesitation, confusion, lost time, and ultimately, lost property and lives. Effective cooperation and coordination of governmental agencies within and among the federal, state, and local governments is essential to achieving a successful homeland security strategy. However, in order to realize resiliency, it is paramount that there is clear cooperation and coordination between the public sector and the private sector.

The importance of the pubic-private initiative is outlined in the Department of Homeland Security's recent National Response Framework (NRF), which defines the roles and responsibilities of the government (federal, state, local, and tribal) and the private sector (private business or NGO). As articulated in the NRF, "Government agencies are responsible for protecting the lives and property of their citizens and promoting their well-being. However, the government does not, and cannot, work alone. In many facets of an incident, the government works with the private-sector groups as partners in emergency management."

The NRF outlines five critical roles played by the private sector during both disasters and terror attacks. First, privately owned critical infrastructures such as transportation, private utilities, financial institutions, and hospitals play a significant role in economic recovery from disaster and terror incidents.[18] Second, "owners and operators of certain regulated facilities or hazardous operation may be legally responsible for preparing for and preventing incidents from occurring and responding to an incident once it occurs."[19] Third, private businesses "provide response resources during an incident—including specialized teams, essential service providers, equipment, and advanced technologies."[20] Fourth, private entities "may serve as partners in local and State emergency preparedness and response organizations and activities."[21] Fifth, private entities play an important role; "as the key element of the national economy, private sector resilience and continuity of operations planning, as well as recovery and restoration from an actual incident, represent essential homeland security activities."[22]

A necessary component to establishing a resilient homeland, there-fore, is a viable public-private sector partnership that is based on (1) defined roles and responsibilities, (2) articulating a coordinated preven-tion-response plan, and (3) repeated training or simulation exercises using the prevention-response plan against realistic disaster/terror scenarios.

Defined Roles and Responsibilities

In forging lasting partnerships between the public and private sectors, the private sector (private business or NGO) must define its role and respon-sibilities relative to the public sector on all government levels (local, state, and federal). Agencies such as the New York Red Cross must work along-side FEMA and the NYPD in an effort to respond to a disaster or another terrorist attack. These partnerships must be created using individual liai-sons to private and public entities predicated on clearly defined roles and responsibilities and open and frequent communication.

Articulating a Plan

The private sector must work closely with the public sector to articulate, develop, and implement a disaster/terror prevention-response plan. Such a plan must implement the clearly defined roles and responsibilities out-lined above. Additionally, a proposed plan need take into account multiple scenarios addressing prevention and response, thereby ensuring that dif-ferent entities are seeking to achieve similar goals. The plan will ensure that different organizations see the big picture and know their particular responsibilities within the larger framework.

Training and Simulation

Fundamental to creating and maintaining the public-private sector ini-tiative is consistent training and simulation exercises. Members of the private and public sectors should conduct scenario-based simulation exer-cises (together and separately) with respect to the proposed plan. These exercises must include realistic disaster scenarios subject to real-life time constraints testing the effectiveness with which both the private and pub-lic sectors respond to complex attacks and disasters. Such training and simulation will ensure that the public and private sectors understand—both theoretically and practically—the vital necessity of cooperation and coordination. Such scenario-based simulation exercises—in highlighting existing institutionalized and systemic weaknesses—most effectively facilitate the development of an effective homeland security strategy.

GOALS FOR PARTNERSHIPS

Public-private partnerships, if properly developed and implemented, are the key to economic recovery. Such a partnership—in the aftermath of a disaster or attack—facilitates the resilience of critical infrastructure, including transportation, utilities, financial institutions, and hospital care. By strategically strengthening security, sharing intelligence, and creating plans for postattack procedures (including having evacuation and transportation plans, identifying places of refuge, and providing basic supplies to aid first responders), such partnerships become the key to a secure and resilient homeland.

Prevention and Resiliency through Intelligence Sharing

The Department of Homeland Security (DHS) has provided excellent guidance regarding how to frame intelligence sharing between the public and private sectors. The importance of information before, during, and after a disaster or attack is vital to resilience. Information sharing is, perhaps, the single most important aspect of successful resilience. Information sharing requires government agencies (federal, state, and local) to share information both among themselves and with the private sector. Furthermore, it requires that the private sector—subject to existing legal and constitutional limits—share information with the public sector. Successful information sharing requires cooperation and coordination both internally (within sectors) and across sectors (between public and private entities).

The process must be institutionalized, requiring a fundamental rearticulation of homeland security strategy. While various public sector agencies are historically hesitant (predicated on policy, culture, and legal restraints) to share information with other agencies—much less the private sector—the lessons of 9/11 and Katrina speak for themselves. Resilience in the aftermath of either disaster or attack requires federal, state, and local government agencies to understand that information sharing is vital to the nation's homeland security. That information sharing process must include the private sector. Otherwise, the mistakes of yesterday will inevitably reoccur.

In addition, the National Infrastructure Advisory Council published a report on private and public sector intelligence coordination and made the following recommendations:

1. Prepare memorandums of understanding and formal coordination agreements describing mechanisms for exchanging information regarding vulnerabilities and risks.
2. Use community policing initiatives, strategies, and tactics to identify suspicious activities related to terrorism.
3. Establish a regional prevention information command center.
4. Coordinate the flow of information regarding infrastructure.

Providing Critical Infrastructure—Continuity Planning

In order to reestablish critical infrastructure after an attack, private entities must have continuity plans. These plans must take into account the known threats, which are only "known" through intelligence sharing between the public and private sectors, as discussed above. These plans must also take into account the components essential to reestablishing the service that the particular entity provides. These plans must provide details regarding how the particular entity will promptly resume service, which may differ depending on the form of attack. In addition, the plan must articulate how the entity will communicate both internally and externally:

1. **Senior executive information sharing:** Develop a voluntary executive-level information sharing process between critical infrastructure CEOs and senior intelligence officers. Begin with a pilot program of volunteer chief executives of one sector, with the goal of expanding to all sectors.
2. **Best practices for the private sector:** The U.S. Attorney General should publish a best practices guide for private sector employers to avoid being in conflict with the law. This guide should clarify legal issues surrounding the apparent conflict between privacy laws and counterterrorism laws involving employees. Moreover, it should clarify the limits of private sector cooperation with the public sector.

3. **Existing mechanisms:** Leverage existing information sharing mechanisms as clearinghouses for information to and from critical infrastructure owners and operators. This takes advantage of the realities that exist sector by sector.

4. **National-level fusion capability:** Establish or modify existing government entities to enable national- and state-level intelligence and information fusion capability focused on critical infrastructure protection (CIP).

5. **Staffing:** Create additional sector specialist positions at the executive and operational levels as applicable in the IC. These specialists should be civil servants who have the ability to develop a deep understanding of their private sector partners.

6. **Training:** Develop an ongoing training and career development program for sector specialists within intelligence agencies.

7. **RFI process:** Develop a formal, and objectively manageable, homeland security intelligence and information requirements process, including requests for information (RFIs). This should include specific, bidirectional processes tailored sector by sector.

8. **Standardize Sensitive but Unclarified (SBU) markings and restrictions:** The federal government should rationalize and standardize the use of SBU markings, especially "For Official Use Only."

COMPARATIVE ANALYSIS

The United Kingdom has enacted legislation requiring contingency plans. That legislation, the Civil Contingencies Act, requires certain private entities to "maintain plans to ensure that they can continue to exercise their functions in the event of an emergency so far as is reasonably practicable."[26] Specifically, entities are required to makes arrangements to warn and inform the public, handle emergencies, and make provisions to ensure that the entity's ordinary functions can be continued to the extent necessary.[27] To ensure effectiveness, the legislation also requires entities enact training programs for those directly involved in the execution of the continuity plan.[28] To assist the entities, the legislation requires local authorities to provide advice and assistance to businesses and voluntary organizations in relation to business continuity.[29]

New York City has taken a first step at creating similar legislation. New York City's Local Law 26 (2004) amended the existing administrative code in relation to building safety in the city.[30] In particular, this new law requires owners of big buildings, in coordination with the FDNY, to prepare detailed plans, train staff members, and conduct full evacuation drills of the entire building every three years.[31] While evacuation plans are an essential first component of a contingency plan, they are not enough to establish even the hope for a resilient homeland. The following is a list of suggested measures that would most effectively facilitate resilience in the aftermath of a disaster or attack:

- Educate the private sector regarding the importance of continuity plans.
- Educate the public about the importance of continuity plans for the private sector.
- Offer expertise in the form of training to enable private entities to create continuity plans.
- Require oversight in exchange for the expertise.
- Pass legislation that puts the private sector on notice regarding the importance of continuity plans.
- Encourage states to pass legislation mandating continuity plans, to the extent a state has such power.
- Offer financial incentives, possibly tax incentives, to entities that establish continuity plans and continue updating those plans.

Conversely, the development of such plans and making them known to the employees and public will engender employee and shareholder confidence. The development and publication of plans contributes to the redevelopment of the economy in the disaster's aftermath. Ignoring these realities leaves corporations vulnerable to significant legal liability. In order to ensure business continuity when employees are prevented from entering a facility, there is a need to address many issues, including computer backups and alternate logistical plans. If employees are ordered to evacuate, corporate leaders must have prearranged plans providing for their safe evacuation, particularly if there is concern that the employees are carrying a foreign agent.

Ignoring the three Ps all but guarantees business failure. Implementing them in the context of risk assessment and business continuity is a basic requirement to minimizing disruption and limiting legal liability. These lead to profits that are, indeed, the ultimate goal of all corporations. By taking the steps now to thoroughly address the three Ps, businesses are able to act affirmatively in protecting their business, employees, and profit. These affirmative steps cost much less in the present than waiting until the attack or disaster occurs.

By example, evacuation of a major city goes beyond telling the public to leave. In the future, officials in New Orleans and elsewhere must plan ahead to ensure the public knows of an evacuation plan and to logistically prepare for a massive evacuation. Evacuation plans, in case of a natural disaster or quarantine, must be directed toward those who cannot fend for themselves. Officials must be educated themselves on what actions they are allowed to take in order to enforce evacuation, quarantine, and other emergency orders.

Evacuation plans for any emergency, hurricane or terrorist attack, must be imaginative and creative—they must include the widespread use of various modes of public transportation, including Greyhound bus lines, school buses, and if need be, buses from outlying municipalities. They must provide for those hospitalized, incarcerated, in nursing homes, and otherwise infirm. Evacuation plans must be developed in cooperation with public associations.

Evacuation plans must include previously designated departure points so that citizens and officials alike know where people can go to be evacuated. Plans must include designated public shelters whose locations are known to the public and are stocked with provisions. Evacuation plans must include sophisticated methods of communicating with the public that go beyond merely informing citizens "you must leave." Evacuation plans must include clearly designated exit routes, and if possible, alternative ones.

Officials must educate the public and themselves. After 9/11, Katrina, Rita, and BP we can no longer say, "This will not happen to us." People must know where they should go in an emergency. Officials need this information as well as where and when to create and enforce quarantine. Officials must prepare for the worst, be it a natural disaster or a terrorist attack. The worst may be yet to come. Discouragingly, business leaders and government officials repeatedly utter the same clichés: "We cannot continue to make the same mistakes." To that end, reports are written, Congress holds hearings, and presidents visit impacted areas vowing decisive action and firm responses. While perhaps calming immediate fears and satisfying a basic need for self-reassurance, the reality is harsh: the overwhelming majority of business leaders—small and large alike— have failed to develop sophisticated business continuity models. There is no justification for this failure; the results of this failure are as clear as the day is long.

157

Implementing—in accordance with industry- and business-specific requirements and subject to economic realities and constraints—the business continuity recommendations proposed in this chapter will significantly contribute to homeland security. Doing so will also directly contribute to asset protection; learning from the chemical plant industry is most effective and instructive.

9

Conclusion—Going Forward

As the title of this book suggests, *the* question with respect to homeland security is: Where are we going? Going forward—this chapter's title—is essential to developing a coherent homeland security strategy. However, the path is not easy primarily because basic terms have not been concisely, precisely, and consistently defined. Furthermore, the effort to create a homeland security policy is hampered by a disconcerting refusal to coherently and honestly identify prevalent threats. Attorney General Holder's stunning congressional testimony,* in which he refused to utter the phrase "radical Islam" in the aftermath of Umar Faisal Shahzad's attempt to blow up a SUV on New York City's 42nd Street, significantly hampers any effort to develop a coherent homeland security strategy.

Simply put, how can there be a policy if decision makers are incapable of clearly articulating what is a known clear and present threat? Whether this extraordinary hesitation is a reflection of political correctness or misbegotten political calculation is both unclear and unimportant; what is essential is that it prevents the development of a coherent homeland security policy, and therefore directly inhibits a "going forward."

Given the extraordinary threats posed, the lack of a concrete proposal as a basis for discussion—if not for practical implementation—is unacceptable. It is as unacceptable as the existing homeland security policy, which, as I suggest throughout this book, must—disturbingly and disconcertingly—be referred to as nonpolicy. Perhaps the more accurate adjective is *lurching* from event to event, reflective of short-term, often tunnel-visioned, thinking.

---

* http://www.youtube.com/watch?v=HOQt_mP6Pgg (last accessed December 20, 2010).

While we have covered some poorly handled incidents and specific agency or response shortcomings in this book, this is neither the time nor place to assign blame for the present state of affairs; that is best left for pundits, commentators, spin masters, and historians. The venomous nature of contemporary American politics and talking head media coverage—marked by raw, unbridled partisanship largely devoid of true substance—significantly hampers any rational-based discussion on critical issues demanding sound, sober, mature discussion. The 20-second soundbite culture has given way to politics best described as a reality show, where lies, gross exaggerations, and "playing with fire" are the norm. Yo-yo policy is counterproductive and can lead to what is, essentially, political whiplash.

The 2010 congressional elections understandably focused, almost exclusively, on the economy; however, as previous events have dramatically demonstrated, homeland security issues will dominate a significant percentage of law and policy makers' time and attention. A continuous election cycle culture, enhanced by the omnipresent blogosphere, facilitates the lack of rational discussion with respect to critical homeland security questions. I contend that our culture of political correctness significantly enhances vapid conversation.

Despite this counterproductive milieu, there is no alternative but to engage in serious discussion regarding homeland security; our future— and our children's—depends on this. As I have discussed throughout this book, the failure to narrowly, precisely, and responsibly define basic terms is an enormous impediment to creating, articulating, and implementing a valid and viable homeland security policy.

Adopting the definition proposed in Chapter 1 will significantly facilitate defining additional concepts relevant to homeland security, including *threat, risk assessment, resource allocation, resource prioritization,* and *cost–benefit analysis.* In addition, defining *homeland security* will enable developing a homeland security policy predicated on long-term strategic thinking rather than endless lurching from event to event, a pattern that largely summarizes the U.S. approach of the past 10 years.

While apologists may correctly remind us that the Department of Homeland Security was created in the immediate aftermath of 9/11 to calm an anxious public and provide concrete security measures, the overarching question is: To what end? Answering that question demands an examination beyond DHS's mission; while that is an important endeavor, our attention must be directed to what are the present and future homeland security threats. Engaging in this important enterprise requires a combination of crystal ball gazing and candid, unvarnished discussion.

As I argue elsewhere, the most significant threat posed to civil democratic society is religious extremism. This is not to suggest other threats do not exist. However, unequivocally, religious extremism demands both vigilance and strategic policy intended to meet the threat head on. The threat is not posed by religion; it is posed by religious extremist actors who are committed to bringing glory and honor to God by randomly—albeit deliberately—attacking innocent civilians. To that threat, homeland security must be directed.

It is, however, not the only threat decision makers must take into account; a quick survey of both successfully executed and foiled terrorist attacks makes that clear. However, terrorist attacks on American soil manifest the extraordinary power of religious extremism and the direct threat it poses to the American public. Perhaps this is an uncomfortable truth; historically, religion has been granted immunity by the public and decision makers.

Developing and implementing an effective homeland security policy requires decision makers, the general public, law enforcement, and people of moderate faith to recognize, internalize, and ultimately proactively act with respect to religious extremism. The responsibility of the general public is to actively engage the issue; denying the threat is not a luxury or an option. To that end, the general public must implore decision makers—policy makers and law enforcement alike—to proactively act against extremist faith leaders actively engaged in inciting their respective communities. I have elsewhere recommended conducting monitoring and surveillance operations against extremist faith leaders known to incite members of their respective congregations. The proposal, admittedly, raises legitimate concerns regarding the chilling effect and limiting freedom of religion and freedom of speech.

However, failure to proactively minimize the extraordinary danger posed by extremist faith leaders reflects both irresponsibility and a failure to look the danger in the eye. That is something that law enforcement and decision makers must recognize and effectively communicate to the public; senior state homeland security officials suggested in a recent conversation that the responsibility lies with executive decision makers. In other words, law enforcement will engage in proactive measures with respect to religious extremism if so ordered. That, frankly, is a convenient response; it is also incorrect, for law enforcement has the responsibility to identify threats and recommend proactive measures to mitigate that specific threat. Hiding behind the discomfort that accompanies a particular threat is simply irresponsible.

With respect to moderate faith leaders, they too must acknowledge the direct threat posed by religious extremists; this requires proactively, publicly, and demonstratively distancing themselves from extremists of their own faith. This demands putting aside possible internal condemnation and castigation; given the perilous threat posed by extremism, narrow political considerations must be rejected. While there is open debate on whether moderates can convince extremists to change their colors, there must be no debate with respect to the need to publicly denounce extremism and the harm caused by extremist actors. When researching my book Freedom from Religion: Rights and National Security in London (December 2008), I met with thoughtful commentators who expressed deep skepticism regarding stated British policy that religious extremists can be influenced by people of moderate faith. I shared their deep skepticism; however, there is undeniable importance in condemnation of extremists by moderates. Unfortunately, this seldom occurs or, at least thus far, not often enough to make a systemic and recognizable impact.

As discussed throughout this book, the traditional U.S. response to attacks on American soil and perceived threats alike has been panic reactions. Without a doubt, the harm caused cuts across many disciplines and borders. From unwarranted violations of personal and civil rights, to ineffective operational decisions, to wasteful spending of resources (financial and otherwise), to creating an illusion of protection and security, to failing to address core issues, the responses all reflect a consistent pattern of failure to create a thoughtful homeland security policy.

There is no need to repeat the litany of responses reflecting the lack of policy over the past 10 years; the critical question is how to articulate, create, and implement an effective homeland security policy. To that end, the ten points below facilitate development of mechanisms necessary to develop such homeland security policy:

1. **Threats:** Narrowly, concisely, and precisely define threats, with the understanding that distinguishing between immediate and secondary threats is mandatory. Not all presumed threats are immediate; not all immediate threats can be equally mitigated. To that end, prioritization and risk assessment are essential; both require intelligence gathering and analysis dependent on developing sophisticated understanding of distinct communities, including sensitivity to particular cultures, mores, and languages.

2. **Realistic expectations:** It is essential that political leadership speak candidly to the public regarding both the threats (see above) and the realistic consequences of resource allocation and prioritization with respect to protecting critical infrastructure. The fundamental requirement is to prioritize resources—honestly and narrowly—predicated on a careful (i.e., not politicized) analysis of intelligence information. The public must be treated respectfully; otherwise, there will be created a dangerous and unrealistic illusion of protection.

3. **Private-public cooperation:** Effective homeland security demands cooperation between government agencies—particularly law enforcement—and the private sector. While law enforcement has traditionally been hesitant to cooperate with the private sector—particularly because of sensitivity relevant to sharing sensitive information—cooperation will facilitate more effective and efficient responses to threats and dangers.

4. **Federal-state-local cooperation:** The 9/11 Commission addressed historical enmity between the FBI and the CIA and the resulting harm to national security; in that spirit, homeland security, in practice and not simply theory, should directly benefit from enhanced cooperation between different branches and levels of government. While responsibility with respect to some issues—public health is an obvious example—is determined by law, that does not justify a lack of cooperation on those issues not statutorily regulated.

5. **Conduct exercises:** All agencies involved in homeland security must commit to regularly conducting exercises that incorporate both training and retraining. Exercises—whether tabletop or field based—are essential to ascertaining levels of competence, professionalism, readiness, and weakness. While exercises require significant financial commitment, their payoff is extraordinary; the almost universal response (including skeptics) to recent simulation exercises is overwhelmingly positive on innumerable levels. Benefits include development of positive working relationships, both internally and with peer organizations, identification of weaknesses in emergency response plans, development of more effective media and public announcements, and enhancement of threat analysis and determination.

6. **Comparative analysis:** Drawing from the experiences of other countries and cultures facing similar threats significantly

enhances developing an effective homeland security policy. While comparison and analogy are imperfect—after all, no two dilemmas and cultures are the same—not drawing on similar experiences is truly a missed opportunity with potential long-term negative consequences. Drawing on professional experience strongly suggests that the commonality of similar experiences has significant mutual benefits that significantly outweigh possible negatives.

7. **Mechanism for implementing security measures:** As these lines are written, air travelers are subject to enhanced pat-downs, intrusive body scans, and additional security measures. Conceivably all measures—individually, in combination, or collectively—significantly enhance the safety of both those traveling and the general public. If that is the case, then arguably the measures are legitimate and demand congressional and public support. However, previous experience has taught the relevant publics to be skeptical when new security measures are introduced; in the immediate aftermath of Northwest Flight 253 TSA announced that TV screens available on flights would be blacked out one hour before scheduled landing, and that blankets and pillows (on those airlines that still carry such basics) would be collected an hour before landing.

Within 48 hours both *new* policies were rescinded, thereby joining the heap of panic responses and justifiably drawing public ridicule for an ill-conceived policy. Needless to say, the most *elementary* of security measures would have prevented Abdulmuttallab from boarding the plane, for *all* the indicators of dangerous passenger were extraordinarily visible: a one-way ticket paid for in cash by a male passenger traveling internationally to Detroit on Christmas Eve without an overcoat or luggage *after* American authorities had been warned by his father regarding his newfound religious extremism. To prevent such an attack, there is no need for enhanced pat-downs or intrusive scanning; there is merely the need for enhanced basic common sense.

While the body scans are unquestionably intrusive, they are defensible provided they make a significant contribution to air safety that cannot be similarly provided by alternative means. While some have suggested that people of faith—in the context of modesty—be exempt from the scans, I find that argument to be without merit. Flying is not a constitutionally guaranteed right; if an individual is uncomfortable with the measures implemented, then alternative travel arrangements can be identified. That, however, is not the issue; the larger, more profound question requires addressing need, alternative, and cost–benefits.

This quote from the TSA director, John Pistole, in response to criticism that travelers should not have to submit to full-body scans or pat-downs—including over-the-clothing contact with the subjects' genital areas—is instructive in that it is general, overbroad, devoid of specific articulated explanation, and leaves more unanswered than answered.

"Clearly it's invasive, it's not comfortable," Pistole said of the scans and pat-downs during an interview on CNN's "State of the Union." But, he added, "If we are to detect terrorists, who have again proven innovative and creative in their design and implementation of bombs that are going to blow up airplanes and kill people, then we have to do something that prevents that."[*]

In other words: Are these intrusive measures really necessary, and what is it they really seek to achieve? Directly addressing those questions would arguably lessen—though not wholly negate—the public outcry that has focused on the intrusiveness rather than on the possible relevance and viability of the measure.

8. **Immigration policy:** Current laws should be reviewed, revised, strengthened, and effectively explained to the general public. Immigration policy is an easily politicized issue, subject to contentious political debate and spins that cut into the core of the American ethos and spirit. President Obama's inability to develop cohesive and coherent immigration reform legislation—campaign promises notwithstanding—reflects both a lack of political will and the reality of the immigration question. Creating an immigration policy is essential to homeland security; while political machinations are inevitable, the reality is that significant decisions have to be made with respect to noncitizens.

Decision makers must resolve the status and rights of individuals presently defined as illegal aliens; the inability to directly address this question impacts individual, public officials, and law

---

[*] http://www.sltrib.com/sltrib/home/50725057-76/tait-tsa-video-shirt.html.csp (last accessed November 22, 2010).

enforcement alike. This conundrum has been at the centerpiece of the immigration debate; failure to resolve the question directly affects homeland security policy from multiple perspectives, including threat and risk assessment and identification, resource prioritization and allocation by law enforcement, and cost–benefit analysis. The importance of the issue is unquestionably magnified because of the increasing danger posed by Mexican drug cartels and their possible influence in and penetration of specific communities in the Southwest.

9. **Bioterrorism threats:** One of the most discussed issues of relevance to bioterrorism is when can public officials order—and law enforcement enforce—a quarantine on an affected or potentially affected population group as a legitimate public health measure. To that end, to prevent disease outbreaks or slow the spread in a bioterrorist attack, the government may enforce vaccinations of individuals and animals alike, regardless of the known dangers.

At what point can we require mandatory testing or destruction of property? These questions come with risks. Potentially, individuals will avoid reporting that they observed an illness among their cattle. In the case of a bioterrorist attack, such as small pox or an illness that affects humans, it is conceivable that people will avoid seeking expensive medical care.

A potential bioterrorism attack requires government officials to determine where to draw the line with respect to limiting individual rights; quarantine policy is but a manifestation of this dilemma in terms of limiting individual rights. Does the law provide sufficient guidelines as to the measures that can and cannot be taken in a bioterrorist attack or other crisis? When exactly can a state of emergency be declared? When is it acceptable to limit civil rights and individual liberties in the name of the greater good? What dangers are involved when we limit rights? Answering these questions is essential to articulating, developing, and implementing a viable homeland security policy addressing possible bioterrorism attacks.

There is a clear need for continued simulations and exercises in order to ensure that all levels of government are prepared for possible attacks. Integrating legal, policy, scientific, and media expertise is essential to minimizing the impact of a bioterrorism attack. In this vein, it is incumbent upon government to proactively educate the public with respect to bioterrorism.

10. **Moving ahead—developing guidelines:** For Congress and the public to determine whether a proposed homeland security measure is effective, the following points represent a suggested matrix:

    - Is the measure preventative, preemptive, or retaliatory?
    - What is acceptable collateral damage?
    - What are the financial costs?
    - What are the costs to civil liberties?
    - What are the geopolitical costs?
    - How valid is the intelligence?
    - To what extent does the measure follow the rule of law?
    - What alternatives exist?
    - To what extent does the measure overlap with existing measures?

Addressing these ten points is essential to policy articulation, development, and implementation. Reasonable minds can reasonably disagree if all the points are essential and how they should be prioritized; however, homeland security demands serious discussion based on threats, resource allocation, and cost–benefit analysis. The ten points directly facilitate the discussion that will answer the question in the book's title: what is homeland security and where is it going?

INDEX